Open Source Software Policy Options for NASA Earth and Space Sciences

Committee on Best Practices for a
Future Open Code Policy for NASA Space Science

Space Studies Board

Division on Engineering and Physical Sciences

A Consensus Study Report of

The National Academies of
SCIENCES · ENGINEERING · MEDICINE

THE NATIONAL ACADEMIES PRESS
Washington, DC
www.nap.edu

THE NATIONAL ACADEMIES PRESS   500 Fifth Street, NW   Washington, DC 20001

This activity was supported by Contract No. NNH17CB02B with the National Aeronautics and Space Administration. Any opinions, findings, conclusions, or recommendations expressed in this publication do not necessarily reflect the views of any organization or agency that provided support for the project.

Cover: Design by Jonathan Lutz, Lloyd V. Berkner Space Policy Intern, Space Studies Board.

International Standard Book Number-13: 978-0-309-48271-4
International Standard Book Number-10: 0-309-48271-2
Digital Object Identifier: https://doi.org/10.17226/25217

Additional copies of this publication are available for sale from the National Academies Press, 500 Fifth Street, NW, Keck 360, Washington, DC 20001; (800) 624-6242 or (202) 334-3313; http://www.nap.edu.

Copyright 2018 by the National Academy of Sciences. All rights reserved.

Printed in the United States of America

Suggested citation: National Academies of Sciences, Engineering, and Medicine. 2018. *Open Source Software Policy Options for NASA Earth and Space Sciences*. Washington, DC: The National Academies Press. https://doi.org/10.17226/25217.

The National Academies of
SCIENCES · ENGINEERING · MEDICINE

The **National Academy of Sciences** was established in 1863 by an Act of Congress, signed by President Lincoln, as a private, nongovernmental institution to advise the nation on issues related to science and technology. Members are elected by their peers for outstanding contributions to research. Dr. Marcia McNutt is president.

The **National Academy of Engineering** was established in 1964 under the charter of the National Academy of Sciences to bring the practices of engineering to advising the nation. Members are elected by their peers for extraordinary contributions to engineering. Dr. C. D. Mote, Jr., is president.

The **National Academy of Medicine** (formerly the Institute of Medicine) was established in 1970 under the charter of the National Academy of Sciences to advise the nation on medical and health issues. Members are elected by their peers for distinguished contributions to medicine and health. Dr. Victor J. Dzau is president.

The three Academies work together as the **National Academies of Sciences, Engineering, and Medicine** to provide independent, objective analysis and advice to the nation and conduct other activities to solve complex problems and inform public policy decisions. The National Academies also encourage education and research, recognize outstanding contributions to knowledge, and increase public understanding in matters of science, engineering, and medicine.

Learn more about the National Academies of Sciences, Engineering, and Medicine at **www.nationalacademies.org**.

The National Academies of
SCIENCES · ENGINEERING · MEDICINE

Consensus Study Reports published by the National Academies of Sciences, Engineering, and Medicine document the evidence-based consensus on the study's statement of task by an authoring committee of experts. Reports typically include findings, conclusions, and recommendations based on information gathered by the committee and the committee's deliberations. Each report has been subjected to a rigorous and independent peer-review process and it represents the position of the National Academies on the statement of task.

Proceedings published by the National Academies of Sciences, Engineering, and Medicine chronicle the presentations and discussions at a workshop, symposium, or other event convened by the National Academies. The statements and opinions contained in proceedings are those of the participants and are not endorsed by other participants, the planning committee, or the National Academies.

For information about other products and activities of the National Academies, please visit www.nationalacademies.org/about/whatwedo.

COMMITTEE ON BEST PRACTICES FOR A FUTURE OPEN CODE POLICY FOR NASA SPACE SCIENCE

CHELLE L. GENTEMANN, Earth and Space Research, *Co-Chair*
MARK A. PARSONS, Rensselaer Polytechnic Institute, *Co-Chair*
LORENA A. BARBA, George Washington University
KELLE L. CRUZ, City University of New York Hunter College
BRENDA J. DIETRICH, NAE,[1] Cornell University
CHRISTOPHER L. FRYER, Los Alamos National Laboratory
JOE GIACALONE, University of Arizona
SARA J. GRAVES, University of Alabama, Huntsville
JOSEPH HARRINGTON, University of Central Florida
ELVA J. JONES, Winston-Salem State University
MARIA M. KUZNETSOVA, NASA Goddard Space Flight Center
CLIFFORD A. LYNCH, Coalition for Networked Information
MELISSA A. McGRATH, SETI Institute
AARON RIDLEY, University of Michigan

Staff

ABIGAIL A. SHEFFER, Senior Program Officer, *Study Director*
NATHAN J. BOLL, Associate Program Officer
ANESIA WILKS, Senior Program Assistant
CARSON BULLOCK, Lloyd V. Berkner Space Policy Intern
JONATHAN LUTZ, Lloyd V. Berkner Space Policy Intern
JACOB ROBERTSON, Lloyd V. Berkner Space Policy Intern

---

[1] Member, National Academy of Engineering.

SPACE STUDIES BOARD

FIONA HARRISON, NAS,[1] California Institute of Technology, *Chair*
JAMES H. CROCKER, NAE,[2] Lockheed Martin Space Systems Company (retired), *Vice Chair*
GREGORY P. ASNER, NAS, Carnegie Institution for Science
JEFF M. BINGHAM, Consultant
ADAM S. BURROWS, NAS, Princeton University
MARY LYNNE DITTMAR, Dittmar Associates, Inc.
JEFF DOZIER, University of California, Santa Barbara
JOSEPH FULLER, JR., Futron Corporation
SARAH GIBSON, National Center for Atmospheric Research
VICTORIA E. HAMILTON, Southwest Research Institute
CHRYSSA KOUVELIOTOU, NAS, The George Washington University
DENNIS P. LETTENMAIER, NAE, University of California, Los Angeles
ROSALY M. LOPES, Jet Propulsion Laboratory
STEPHEN J. MACKWELL, Universities Space Research Association
DAVID J. McCOMAS, Princeton University
LARRY PAXTON, Johns Hopkins University, Applied Physics Laboratory
ELIOT QUATAERT, University of California, Berkeley
BARBARA SHERWOOD LOLLAR, University of Toronto
HARLAN E. SPENCE, University of New Hampshire
MARK THIEMENS, NAS, University of California, San Diego
ERIKA WAGNER, Blue Origin
PAUL WOOSTER, Space Exploration Technologies
EDWARD L. WRIGHT, NAS, University of California, Los Angeles

Staff

COLLEEN HARTMAN, Director (after August 6, 2018)
MICHAEL H. MOLONEY, Director (until March 2, 2018)
RICHARD ROWBERG, Interim Director (March 2, 2018 to August 6, 2018)
CARMELA J. CHAMBERLAIN, Administrative Coordinator (until June 30, 2018)
TANJA PILZAK, Manager, Program Operations
CELESTE A. NAYLOR, Information Management Associate
MARGARET A. KNEMEYER, Financial Officer

---

[1] Member, National Academy of Sciences.
[2] Member, National Academy of Engineering.

Preface

The Committee on Best Practices for a Future Open Code Policy for NASA Space Science of the National Academies of Sciences, Engineering, and Medicine was charged to investigate and recommend best practices for the National Aeronautics and Space Administration (NASA) as it considers whether to establish an open code and open models policy, complementary to its current open data policy. The committee's complete statement of task is reprinted in Appendix A.

To address its task, the committee worked with a lawyer who specializes in open source software licensing and intellectual property rights as an unpaid consultant, held three in-person meetings and many teleconferences during its work from October 2017 through August 2018, and solicited community input via white papers and presentations. The meetings included extensive conversations with NASA leadership from diverse areas within the organization, including the Science Mission Directorate (SMD), the Space Technology Mission Directorate (STMD), the Office of the Chief Scientist (OCS), and the Office of the General Counsel (OGC), as well as with policymakers from other government agencies. The committee also received presentations from a broad range of stakeholders, including researchers across the SMD disciplinary communities, leading experts in computer science and open source architectures, representatives from academic journals and publishing organizations, and the lead author of a concurrent National Academies advisory report on open science. The committee's broad call for white papers was primarily targeted at the SMD disciplinary communities but was open to anyone who wanted to provide input to the study process. The white paper call and listing of received papers is reprinted in Appendix C.

The committee was careful to remain within the scope of its task by defining the complex issues and policy options that NASA will need to consider when deciding to implement a future open code policy, while avoiding any recommendations as to whether or not NASA should implement such a policy. Chapter 1 of this report describes the motivation, goals, and processes undertaken during the study. Chapter 2 provides fundamental background materials, such as the definitions of common terminology, references to relevant legal statutes, and information about open source software as a licensing model and as a development model. Chapter 3 describes the past and current states of software and data management policies at NASA and other related government institutions. Chapter 4 delineates the lessons learned from prior experience with open source software, aggregated from community input. Chapter 5 presents a series of policy options identified by the committee that reflect the choices NASA will need to make in balancing the competing needs of stakeholders while meeting a variety of legal obligations, often conflicting, summed up in the maxim, "as open as possible, as closed as necessary." Chapter 6 presents a summary discussion.

The committee would like to thank the many generous individuals at NASA and other U.S. government agencies and within the greater scientific community who contributed to the study process through presentations, written input, and discussions. A special thanks goes to the staff of the Space Studies Board—Abigail Sheffer, Nathan Boll, Anesia Wilks, Richard Rowberg (interim director), and former director Michael Moloney. Finally, the committee would like to acknowledge and thank Diane Peters, general counsel at Creative Commons, for the invaluable legal insight and expertise she provided throughout this study.

Acknowledgment of Reviewers

This Consensus Study Report was reviewed in draft form by individuals chosen for their diverse perspectives and technical expertise. The purpose of this independent review is to provide candid and critical comments that will assist the National Academies of Sciences, Engineering, and Medicine in making each published report as sound as possible and to ensure that it meets the institutional standards for quality, objectivity, evidence, and responsiveness to the study charge. The review comments and draft manuscript remain confidential to protect the integrity of the deliberative process.

We thank the following individuals for their review of this report:

Christine L. Borgman, University of California, Los Angeles,
Adam S. Burrows, NAS,[1] Princeton University,
Mark Cheung, Lockheed Martin Solar and Astrophysics Laboratory,
Steven D. Christe, NASA Goddard Space Flight Center,
Eric Dashofy, The Aerospace Corporation,
Thomas A. Kalil, Schmidt Futures,
Julianne I. Moses, Space Science Institute, and
Sharon Woods, Department of Defense.

Although the reviewers listed above provided many constructive comments and suggestions, they were not asked to endorse the conclusions or recommendations of this report nor did they see the final draft before its release. The review of this report was overseen by Robert F. Sproull, NAE,[2] University of Massachusetts, Amherst. He was responsible for making certain that an independent examination of this report was carried out in accordance with the standards of the National Academies and that all review comments were carefully considered. Responsibility for the final content rests entirely with the authoring committee and the National Academies.

---

[1] Member, National Academy of Sciences.
[2] Member, National Academy of Engineering.

Contents

SUMMARY	1
1 INTRODUCTION AND POLICY PURPOSE	10
1.1 History and Motivation, 10	
1.2 Policy Goals, 13	
1.3 Committee Process, 14	
2 BACKGROUND MATERIALS	15
2.1 Definitions, 15	
2.2 Categories of Software, 16	
2.3 Legal Issues, 18	
2.4 Licenses—Spectrum of Openness, 22	
2.5 Open Source as a Development Model, 25	
3 PAST AND CURRENT POLICIES	28
3.1 Data Policies, 28	
3.2 Data Management Plans, 30	
3.3 Software Policies, 34	
3.4 Journal Policies on Open Data and Software, 46	
4 LESSONS LEARNED FROM COMMUNITY PERSPECTIVES	48
4.1 Impact of Open Code, 49	
4.2 Education and Training Needs, 53	
4.3 Funding and Effort Needs, 55	
4.4 Enabling Credit and Career Advancement, 58	
5 POLICY OPTIONS AND RECOMMENDATIONS	60
5.1 Policy Option A: Continue Status Quo, 61	
5.2 Policy Option B: Incentivize Openness to Accelerate the Change, 61	

    5.3 Policy Option C: Mandate Openness, 65
    5.4 Transitioning Toward Openness, 66
    5.5 Assessment and Future Considerations, 69
    5.6 Policy Implementation, 70

6   DISCUSSION   73

APPENDIXES

A	Statement of Task	79
B	Copyright Issues of Interest to NASA Investigators and Developers of Software	80
C	Call for White Papers and Listing of Received White Papers	82
D	Biographies of Committee Members and Staff	87
E	Acronyms	93

Summary

INTRODUCTION

Modern science is ever more driven by computations and simulations. In particular, the state of the art in space and Earth science often arises from complex simulations of climate, space weather, and astronomical phenomena. At the same time, scientific work requires data processing, presentation, and analysis through broadly available proprietary and community software.[1] Implicitly or explicitly, software is central to science. Scientific discovery, understanding, validation, and interpretation are all enhanced by access to the source code of the software used by scientists.

This report of the Committee on Best Practices for a Future Open Code Policy for NASA Space Science of the National Academies of Sciences, Engineering, and Medicine investigates and recommends options for NASA's Science Mission Directorate (SMD) as it considers how to establish a policy regarding open source software (OSS) to complement its existing policy on open data. In particular, the report reviews existing data and software policies and the lessons learned from the implementation of those policies, summarizes community perspectives, and presents policy options and recommendations for implementing an OSS policy for NASA SMD.[2]

The purposes of any open code policies that SMD may develop are to serve the goals of SMD and NASA. Based on NASA's vision, mission, and mandates; guidance provided to the study committee from NASA representatives; and the general interests of science and society, the committee identified the following seven goals:

1. Enhance and enable innovation and discovery.
2. Increase the visibility, accessibility, and reuse of NASA-funded code.[3]
3. Facilitate scientific reproducibility.

---

[1] *Community software*, also called community source software, consists of software developed by a group effort, usually in an open source environment. These software packages are often widely reused.

[2] This report does not address software related to spacecraft operations, because of safety and national security concerns that would be raised by the publication of such software. Neither does this report deal with software that supports NASA's administrative functions. Rather, the report focuses on software related to NASA's core science interests: research and technology development related to basic science and engineering, design concepts, mission support, aeronautic vehicles, and major research and engineering facilities.

[3] *Source code* consists of sets of human-readable statements written in a programming language that together compose software. The terms *code* and *source code* are often used interchangeably. *Software* is a general term used for computer programs and applications that provide users with some degree of utility or produce a result or service. Software can be distributed in executable form, as source code or as a service via the Internet.

4. Encourage collaboration inside and outside NASA.
5. Maximize NASA's benefit to society.
6. Respect the security and privacy of citizens.
7. Comply with broader government policies.

A guiding maxim behind all of these goals—and the recommendations in this report—is that software needs to be as open as possible; as closed as necessary.

Past and Current Policies

The committee reviewed data policies at NASA and the U.S. Geological Survey (USGS); data management plans at NASA, the National Science Foundation (NSF), and USGS; software management at a number of federal agencies; federal policy; and large community software projects. Many of the lessons learned can be summarized by the statement "Software is data, but data is not software." Software is included in the definition of data (Section 2.1), but software is copyrightable, whereas data is not. The ability to claim copyright is an important distinction in software versus data policy implementation. Yet, there are important lessons to be learned from the implementation of open data policies.

Some fields within NASA and elsewhere have already established a culture where openness is expected. An open data access policy for Landsat satellites, which are jointly managed by USGS and NASA, has dramatically increased the economic value and exploitation of the data for research, commercial, and land use management applications. NASA's efforts to share data have been facilitated by the development of infrastructure, such as formal data archive centers, by SMD. These centers enable responsiveness and timely delivery of data to the public, facilitate data archiving, and provide data access and visualization tools that would be inefficient for individual researchers to create.

The expansion of science enabled by making data open to the public has driven efforts for even greater openness across the federal government (see Section 1.1). Today, federal agencies are required to release at least 20 percent of new software as open source, and NASA encourages vendors to use open source technology.[4] Indeed, it is likely that success in the development and implementation of open data plans will continue to create an environment that fosters the development and implementation of open software. While open access policies can dramatically increase the economic value and exploitation of federally funded resources and have unanticipated applications that benefit society, both the policy review and lessons learned from community perspectives highlight the need for a careful and thoughtful process that responds to community feedback during the transition to any new policy. The committee found that it was the changes in agency data policies that prompted changes in community norms such as accepted practices regarding data sharing. Community understanding of new requirements for open data, and prospectively for OSS, has been facilitated by the use of consistent language, clear proposal submission guidelines, and the availability of educational resources.

Software comes in many varieties, and developing open source policies and implementation plans for software is more complex than for data. In some cases, software policy will need to take into account funding sources, the parties involved, the development history, and the size and complexity of the software and computational requirements. For example, a large community-source software project may involve contributions from different institutions, agencies, and countries, each of which may have its own software policies and legal constraints that impede software sharing, export, and licensing practices. Crafting a workable policy in such situations will not be easy. Nonetheless, there are several practices that could generally build support for and facilitate the implementation of OSS policies.

Some scientific communities have already embraced OSS, but others are less familiar with it. Within NASA, program managers understand their research communities best and are in close contact with many of the scientists that new policies would impact. Enacting any new policy that requires a shift in culture also will require community

---

[4] This Office of Management and Budget (OMB) directive was intended to be a 3-year pilot, but it appears to still be in effect and was referenced in the Environmental Protection Agency (EPA) 2018 policy, https://www.epa.gov/open/interim-open-source-software-oss-policy.

support for successful and efficient implementation. Building support for OSS could include pilot studies, gaining the support of program managers, the use of funding as an incentive for researchers, clear reporting guidelines and evaluation criteria, and evaluation of compliance by groups that are independent from a given program—to increase the confidence of the community that open source policies are being fairly implemented. The adoption of open source policies is also being facilitated by journals and publishers who are moving forward in support of a more open science environment, providing both enhanced recognition and access to data and software. An initial and continued assessment of NASA-funded research projects for current OSS practices would be useful for analysis and fine-turning of policy implementation. Understanding community practices sets a baseline for future progress.

Community Perspectives

The committee examined the perspectives of NASA's space and Earth science community through the collection of white papers from and conversations with members of the community (see Appendix C). Overall, the community expressed broad support for OSS. Openness and transparency are seen as central to scientific validity and reproducibility, but various challenges occur in the implementation of policy. A majority expressed positive experiences in opening code and described a range of advantages, including efficiency, greater collaboration, reduced duplication, greater use of particular codes, more robust code, and a broadening of the user community. OSS can improve the testing of codes, facilitate the ability of scientists to conduct reproducible research, and enhance the transparency of research.

Many of the white papers, on the other hand, emphasized issues and even pitfalls when trying to regulate the open sourcing of software. Concerns included legal ramifications, institutional barriers, costs, the level of effort required to implement OSS policies, the need for training and education, and other impacts on individual scientists and their careers. Some suggested that an open source policy may not always benefit science, because for researchers, time spent publishing software comes at the expense of time spent doing science. While an open source policy may enhance science for other researchers, it could be at the expense of the original researcher's scientific output. In addition, there are concerns that researchers may lose motivation to push the boundaries of innovation in their software if they know that they have to immediately release it to the general public instead of having several years to take advantage of the new technology, potentially leading to less innovation in software development. Because doing science and developing OSS are different, but complementary, activities with different motivations and outcomes, OSS policies may be more successful if they clearly identify value in both activities.

Many concerns reflect misunderstandings about open source licensing and processes. Others reveal legitimate legal and institutional barriers. In some cases, it will be necessary to change the culture of institutions as well as scientists. Most unease within the community stems from the culture of how science is currently competed and conducted. For many software projects, open sourcing the code from inception is ideal. For others, a period to verify and validate the code in a research mode may be a better approach. SMD will need to address these concerns as it develops open source policies. In particular, SMD needs to foster a new culture of openness and encourage a social norm of sharing and collaboration, in part by incentivizing the development of OSS in the academic community through the use of targeted grants, fellowships, and prizes. The move toward openness is also facilitated by the establishment and use of open source libraries (code and tools used by programmers when writing software) to collect and disseminate community software. An incremental and flexible approach (as discussed in the policy options below) to OSS software will allow researchers to adjust to new requirements and minimize the impact on their scientific productivity.

Work toward a cultural norm of openness has already begun, with the establishment of open data policies, support, and infrastructure. This progress needs to continue with carefully constructed support for OSS, beyond simple policy development and implementation. A well-defined open source policy, combined with functional processes for review and release of software, can substantially reduce fear, uncertainty, and doubt about making software open source and can be a major enabler of open source development. The finding and recommendations below are based on the committee's assessment of community perspectives.

Finding: The NASA science community generally recognizes the value of open source software and supports the principles of openness, but concerns prevail on the details of implementation and the impact on science and scientific careers.

Recommendation: NASA Science Mission Directorate should explicitly recognize the scientific value of open source software and incentivize its development and support, with the goal that open source science software becomes routine scientific practice. (Chapter 4)

Recommendation: NASA Science Mission Directorate should initiate and sponsor programs to educate and train researchers in open source best practices. Topics could include, but are not limited to, export controls, licensing and intellectual property, workflows, and software development. These resources could be made available to the community via in-person trainings as well as webpages, screencasts, and webinars. (Chapter 4)

Recommendation: Any open source software policy that NASA Science Mission Directorate develops should not impose an undue burden on researchers; therefore, any policy should be as simple as possible, and any mandates should be fully funded. (Chapter 4)

Recommendation: NASA Science Mission Directorate should support the infrastructure, governance, and maintenance of a healthy open source community, taking advantage of existing community resources to the greatest extent possible. (Chapter 4)

Recommendation: NASA Science Mission Directorate should support open source community-developed libraries that advance NASA science. (Chapter 4)

Recommendation: NASA Science Mission Directorate should foster career credit for scientific software development by encouraging publications, citations, and other recognition of software created as part of NASA-funded research. (Chapter 4)

Policy Options

In this section, the committee outlines a selection of policy options, including both incentives and mandates, for NASA SMD to consider. Based on the charge to the committee and discussion with NASA officials, the committee operated under the assumption that SMD will transition to a greater level of openness in accordance with federal policy. It is important, therefore, that NASA ensures that the transition helps advance science, foster collaboration, and generally advance the goals listed above. The committee believes that the best way to achieve this is to work toward a cultural norm of robust OSS development and maintenance. This will not happen overnight and will require ongoing strategic investment.

The options below can be considered a sort of toolbox to draw from to help move the community toward greater openness while recognizing that different disciplines and code types will have different requirements and transition at different rates. Incentives will help to move the community norms toward greater openness regardless of whether mandates are eventually implemented. Overall, the committee believes that there will need to be a combination of different incentives in place and transition to mandates only as appropriate.

Recommendation: NASA Science Mission Directorate should consider a variety of policy options depending on discipline and software type and transition to greater openness over time. (Chapter 5)

The committee identified the following three OSS policy options:

Option A: Continue status quo.
Option B: Incentivize openness.
Option C: Mandate openness.

Policy Option A: Continue Status Quo

Currently, NASA SMD has no division-wide OSS policy regarding software publishing, distribution, or licensing. Option A would continue to allow individual NASA programs to determine whether they and their research communities are interested in moving toward open source. Some programs and modeling centers have already taken steps toward openness. Option A could eventually lead to OSS being required or becoming a de facto norm in some areas, but in others it would remain unusual. Without SMD coordination of an OSS policy, missteps in one division could potentially be repeated in another.

Policy Option B: Incentivize Openness

Option B would preserve community interests while gradually moving to the wide adoption of OSS. The goal of this option is to build trust while working toward making openness a community norm. With mandates absent or delayed, community pressure toward openness would naturally increase as investigators compete for the incentives.

Success with this policy option would depend on the allocation of adequate resources. Incentives within the current budget that lead to reduction of research funds will be less accepted by the community. There may be a delay in the scientific return from research funding. Over the long term, however, moving toward more openness is likely to provide a net benefit to science, as more researchers take advantage of open software. However, because incentives to adopt OSS may be applied at different rates and perhaps be absent altogether at some governmental agencies, some researchers may not participate, possibly gaining a research or career advantage over those who devote time and resources to opening their software.

The committee identified five specific elements, one or more of which could be adopted as part of Option B.

B1. Funding for new proposals specifically addressing an OSS need
B2. Funding augmentations or components of proposals to open and support software
B3. Piloting the use of software management plans in some programs
B4. Supporting open source libraries and infrastructure software development
B5. Offering a prize for exemplary contributions to OSS in the NASA science community

One or more of these elements could be adopted as part of Option B. Each element has its pros and cons, and they are likely to be applied differently for different software types.

Option B1—Funding for Full Open Source Software Proposals

Under Option B1, SMD or its divisions would allocate funding for new proposals addressing an OSS need, such as to open existing software with community reuse potential or replace it with functionally equivalent OSS, to develop new or maintain existing OSS, or to extend community open source libraries and frameworks. Proposals would be required to include a software management plan (SMP) that describes new software produced during a project and how it will be handled during the project and archived afterward. This option allows for a prioritized approach that recognizes the cost of community software development and creates or builds scientific software projects that other researchers can reuse.

This option could delay scientific returns within the programs implementing it, as scientists spend time to gain experience and familiarity releasing software. It would open only some software, and it may provide a disincentive for groups who do not win funding to open their software.

Option B2—Optional Proposal Open Source Add-On

Under Option B2, scientific research proposals submitted to SMD in existing grants programs could optionally include a distinct section to justify additional effort and funding to open software from the project and to provide a software management plan. Unlike B1, the OSS management plan and funding is an augmentation to the scientific proposal. This option is otherwise similar to B1, with the difference that there may be situations where the scientific merit of a proposal is not rated high enough for support, but the open source add-on is seen to have significant value.

Option B3—Pilot Software Management Plans

Under Option B3, specific programs within SMD would begin to require SMPs for scientific proposals containing substantial new software development as part of the proposed research. Requiring an SMP would not mandate software openness, but it could gradually expand existing policy and impose more specific requirements over time. In addition, the requirement for SMPs would be phased in; initially, it would apply only to selected SMD programs. The goal would be to gradually develop an effective policy by identifying different approaches to making software more open and by responding to community feedback. The gradual implementation of requirements for using SMPs would reduce the extent to which such requirements would disrupt SMD programs, and it would allow SMP requirements to be fine-tuned based on the results of the pilot programs.

However, Option B3 imposes an additional requirement that researchers must adhere to and that evaluators must consider. If B3 is implemented rapidly in scientific communities unfamiliar with OSS, innovation and science could suffer due either to inexperience making software open or through the additional burden of time spent by researchers on software. It is unclear what implications successful proposers would face if their stated SMP goals were unmet, except by evaluating previous practices, which would apply only to previous OSS funding from NASA.

Option B4—Support for Open Source Libraries and Infrastructure Software

Under Option B4, SMD would use existing funding mechanisms or allocate SMD employees to support and adopt open-source libraries and infrastructure software that are widely used in NASA-funded research. This option could improve community software quality and generate savings for NASA as a whole. Some software of this type, however, currently exists without dedicated NASA funding.

Option B5—Annual Prizes for the "Advancement of Open Source Software Development and Impact"

Greater recognition for scientists for creating quality OSS would enhance their career advancement. Under Option B5, an SMD award or prize could provide some recognition and visibility for the importance of OSS. This prize would recognize how OSS provides value to NASA. Implementing this option, however, would require resources and take time away from other activities, because SMD needs to create the prize, publicize it, organize a review committee, review applications, and make a selection. Awards and prizes are not nearly as effective at advancing scientific careers as funded proposals unless they are extremely prestigious.

Policy Option C: Mandate Openness

Under Option C, NASA SMD would decide that, by a certain date, software created through NASA SMD funding will be open source, with only a few, strongly justified exceptions. Mitigating the concerns raised in Chapters 3 and 4 would require a period of transitional activities, happening at rates that may vary by program and software type. Moving too rapidly into a mandate would likely be counterproductive and jeopardize future OSS transitioning efforts. The transition would require resources for training, software support and maintenance, and contributions to the overall software infrastructure. A mandate would be the surest and quickest way to increase the transparency of NASA science and to satisfy relevant federal policies; it could potentially enhance NASA's national and international reputation as a leader in open science; and experience with open data policies suggests that an OSS mandate could drive other agencies, both nationally and internationally, to enact similar policies, thereby benefiting NASA SMD researchers. A mandate, however, may be the costliest option, requiring major enabling and sustaining infrastructure to enforce the mandate. For some software types, the cost could exceed the benefit. A mandate could flood repository sites with a large variety of software. A mandate might also hinder collaboration with other agencies in creating OSS, notably the Department of Defense (DOD), which would have to provide permission and may have additional security concerns to protect controlled unclassified information and export-controlled information. Mandates are more likely to be effective once incentives have been established and only if mandates are implemented over a carefully planned flexible transition period.

Conclusion: Immediately mandating open source across all software types and in all of SMD could damage the NASA science enterprise.

Conclusion: An incentive-driven transition period is needed before a comprehensive SMD open source software policy. Incentives and timelines will vary by software type and community experience.

Transitioning Toward Openness

A variety of policy options, which will depend on discipline and software type, are warranted and will facilitate the transition to greater openness over time with a clear path toward openness. In Chapter 5 (Section 5.4), the committee describes in detail which policy options may be appropriate for each of seven defined software types and a suggested time frame for moving to mandated openness. The committee considers 3 years to be the minimum transition time, which is applicable to only some software types or communities. Many programs or software types will transition more slowly because of different grant cycles, infrastructure availability, and general community readiness—although they will follow the same general path. There will, however, be limitations—some software cannot legally be open source, some legacy software may simply be too expensive to convert, and different software will have different levels of maintenance (sometimes none). SMD will need to continually balance trade-offs and priorities while continually assessing how policies are meeting their goals. This transition to a desired level of openness requires time and resources for training, software support and maintenance, and contributions to the overall software infrastructure. Introducing OSS requirements without strategic investment in software development and maintenance may not advance innovation and discovery and other goals.

An assessment of how the scientific community uses software before, during, and after implementation of policies could help advance policy goals more efficiently. Implementing a policy that affects how scientists perform their research is a delicate undertaking, and continual assessment will be important to minimize disruptions and capitalize on successes. A measure of how changes in software release policies relate to scientific efficiency and advancement may be difficult to clearly articulate, especially in a short time frame, since many effects related to shifts in policy may develop slowly. Different measures may be needed in different communities. Nevertheless, some attempt at assessment would likely improve policy implementation.

Policy Implementation

Licensing

The Copyright Act of 1976 ensures that any original creative work, including computer source code, is automatically protected by copyright once created, except for work created by the federal government, which is excluded by U.S. Code.[5] This generally restricts use of software unless the owner has granted a license. A license is considered a public license if the owner grants permissions to the public as a whole to use their software as long as they abide by the license terms.

---

[5] 17 U.S. Code § 105 - Subject matter of copyright: United States Government works.

Some licenses are more permissive than others. The most permissive licenses place the least restrictive conditions on use and are very close to dedicating the software to the public domain, putting few restrictions on use or modification of the software for any purpose. This allows users to restrict the redistribution of their own software and contributions by others, even if that software is derived from OSS. In contrast, the most restrictive class of open source licenses, exemplified by GPL,[6] requires that derivative works, if released, be released under the same open source license as the original. This ensures improvements and changes to OSS are shared back with the public. To be considered open source, software needs a license that complies with the Open Source Definition. One of the criteria of the definition is that open source licenses "must allow modifications and derived works, and must allow them to be distributed under the same terms as the license of the original software."[7]

NASA SMD releases some software created by civil servants under the NASA Open Source Agreement (NOSA 1.3).[8] The Open Source Initiative (OSI) approved NOSA, but some controversy about specific provisions in the agreement subsequently ensued, and it was determined to be incompatible with GPL.[9] NASA indicated that it has attempted to address some of these compatibility concerns in its latest version, NOSA 2.0, but approval of NOSA 2.0 by OSI remains pending as of the date of this report. NASA will need to consider how best to balance the different goals of enabling innovation, facilitating scientific reproducibility, stimulating the economy, and benefiting society when recommending particular licenses that are as open and permissive as possible and only as closed as necessary.

Recommendation: NASA Science Mission Directorate should encourage the use of standard open source licenses, but not mandate a particular license. Nonstandard licenses should be justified in the software management plan. (Chapter 5)

Software Release

Software released by NASA employees undergoes a rigorous vetting process to ensure the legality of its release, ensure compliance with software engineering standards, and prevent disclosure of restricted information. The same process applies for all software regardless of the length, topic, or a priori risk of the code. The NASA Technology Transfer Program has recently made major improvements to the process. Nevertheless, NASA's current internal software release policy procedures can cause undue and potentially harmful delays in the release of low-risk software. The process could also be improved for software in more sensitive areas by identifying the likely risks, working with legal experts in planning software to reduce risk, and expediting the final review by focusing it on the areas of concern.

Recommendation: NASA Science Mission Directorate should develop internal policies and external legal language conducive to the swift release of open source scientific software, and the full participation of NASA employees in internal and external open source projects, without jeopardizing national security or incurring legal liability. (Chapter 5)

Ongoing Compliance

Technology changes quickly, and this affects how scientists do research and create software. Any OSS policy based purely on licensing considerations could possibly be circumvented. For example, software as a service (SaaS) is a delivery model where users access software and data through a web interface. Since the software itself is not copied, copyright license terms may not be triggered, particularly for permissive open source licenses. This raises concerns about source code access, reproducibility of science, and long-term sustainability and maintenance,

---

[6] GNU General Public License (GPL). More information at https://www.gnu.org/licenses/gpl-3.0.en.html.
[7] See https://opensource.org/licenses.
[8] See https://opensource.org/licenses/NASA-1.3.
[9] See https://en.wikipedia.org/wiki/NASA_Open_Source_Agreement.

because the availability of the software may change without notice to the user or disappear entirely during an investigation. So even though SaaS and other computing technologies can be used in a positive way, they can also be used as a mechanism to circumvent policy, and software management plans will need to address the use of SaaS in new software development. However, there is at least one open source license that includes provisions prohibiting use of the licensed code as SaaS.

Discussion

When it comes to software, openness is not enough to advance science. The basic act of releasing software as open source is not difficult, but it can evoke some complex considerations. To make OSS truly useful and helpful in realizing NASA's goals requires a coordinated, end-to-end development approach supported by infrastructure, community practices, and education over the long term. Ultimately, OSS will likely advance science, but there will be transition and maintenance costs requiring a careful balance of trade-offs and active engagement by program managers.

The recommendations in this report cover options to increase community education and training and to ease implementation of new policies or requirements. It is important to note that most of the committee's recommendations apply regardless of whether NASA SMD explicitly requires OSS. Creating a cultural shift toward greater openness will be challenging. Many of the lessons learned from the implementation of open data policies can be applied to the implementation of an OSS policy; however, OSS is more complex than open data. The recommendations allow for an implementation of open source that is carefully planned, gradually implemented, and well-coordinated across NASA SMD.

1

Introduction and Policy Purpose

Modern science is driven by computations and simulations. Discoveries and insight often come from complex simulations of climate, space weather, and astronomical phenomena. At the same time, scientific work requires regular data processing, presentation, and analysis through broadly available proprietary and community software. Implicitly or explicitly, software is central to science. Scientific discovery, understanding, validation, and interpretation are all enhanced by access to the source code of the software used by scientists.

1.1 HISTORY AND MOTIVATION

The National Aeronautics and Space Act of 1958 created NASA as an organization whose first goal is the "expansion of human knowledge of phenomena in the atmosphere and space."[1] The act also directs NASA to "provide for the widest practicable and appropriate dissemination of information concerning its activities and the results thereof." This mandate underpins NASA's mission, and NASA has a long history of encouraging openness in the research it conducts and sponsors.

Before the 1990s, digital data sharing was cumbersome, involving mailed magnetic tapes, compact disks, or hard drives. The scientist who physically held the data-storage medium controlled access, thereby limiting scientific advancement and reproducibility of results. With the advent of inexpensive digital storage and fast transfer of information over the Internet, it became easier to share data, and agency policies began adapting. In 1994, the National Oceanic and Atmospheric Administration (NOAA) and NASA Earth Science Division (ESD) committed to a full and open data policy for all civil Earth observation satellites.[2]

Following the movement toward open data is a movement to open software. The 2000 report of the President's Information Technology Advisory Committee, *Developing Open Source Software to Advance High End Computing*, recommended that the "Federal Government should encourage the development of open source software as an alternate path for software development for high end computing."[3] It also recommended an analysis of existing open source licenses that could be distributed to various agencies, and that "the use of common licensing agreements should be encouraged."

---

[1] See https://history.nasa.gov/spaceact.html.
[2] See https://earthdata.nasa.gov/nasa-data-policy.
[3] President's Information Technology Advisory Committee, 2000, *Developing Open Source Software to Advance High End Computing*, Report to the President, https://www.nitrd.gov/Pubs/pitac/pres-oss-11sep00.pdf.

NASA later published the report *Developing an Open Source Option for NASA Software*, which states in the introduction, "Open Source is about enhanced software quality, more efficient software development, and increased collaboration."[4] It also acknowledges that the Open Source Initiative (OSI) "provides the most widely recognized guidelines as to what constitutes open source." The report reviews the leading open source licenses (all OSI-approved) and associated issues such as export controls, the directions in "External Release of NASA Software" (NASA Procedures and Guidelines [NPG] 2210.1A), contractor rights, and copyright for software created by government employees. In particular, it highlights Section 3.4.3.2.2 of the NPG, which states that "software that is joint work between NASA employees and NASA contractors is protected under copyright." It finally proposes that NASA utilize the Mozilla Public License (MPL), avoiding the "need to develop yet another license and submit it to the OSI for approval." The committee will consider these issues in Chapter 2.

NASA decided to develop a new license, the NASA Open Source Agreement (NOSA), specifically designed for software generated by civil servants, which was approved by the OSI.[5] NOSA has not been widely accepted by the open source community due to what appears to be lack of understanding about the need for some of its unique provisions and differing interpretations about those same provisions, among other concerns. Chapter 2 will expand on issues associated with NOSA and review open source licensing in general.

To better understand the OSS process, NASA held three open source summits, in 2011, 2012, and 2013 to discuss how to improve the development and release of OSS at NASA, how to advance the use of OSS throughout a wider government audience, and how to engage with and encourage open source communities, respectively.[6] Also, the website open.nasa.gov was created to promote the agency's OSS, data, and application programming interfaces.

Meanwhile, the expansion of science enabled by making data open to the public has influenced a drive for even greater openness across the federal government. In 2013, the Office of Science and Technology Policy (OSTP) issued the memorandum "Increasing Access to the Results of Federally Funded Scientific Research."[7] The memo commits each research and development agency to ensure that "the direct results of federally funded scientific research are made available to and useful for the public, industry, and the scientific community. Such results include peer-reviewed publications and digital data." Furthermore, it directs each agency to "develop a plan to support increased public access to the results of research funded by the Federal Government." In response to the OSTP memo, in 2015 NASA developed a *Plan for Increasing Access to the Results of Scientific Research*, which addresses data, but not software.[8]

In 2016, the Office of Management and Budget (OMB) issued the memorandum "Federal Source Code Policy: Achieving Efficiency, Transparency, and Innovation Through Reusable and Open Source Software."[9] This memo requires agencies to release at least 20 percent of new custom-developed software as open source, meaning that the source code is available and licensed for reuse, to increase efficiency across the federal government. The NASA Office of the Chief Information Officer (OCIO) responded that beginning in 2017, NASA will comply with the requirement and also when contracting for software development, "NASA will encourage vendors to use open source technology wherever possible."[10] These policies are specific to development of information technology and software solutions and do not include any directives regarding scientific research software. Different scientific disciplines, even within NASA, have a variety of experience, familiarity, and comfort sharing data, models, and software. Some fields have already established a culture where openness is expected (see discussion in Section 3.3.1).

It is in this context that NASA's Science Mission Directorate (SMD) requested that the National Academies of Sciences, Engineering, and Medicine conduct a study on "Best Practices for a Future Open Code Policy for

---

[4] Moran, P.J., 2003, "Developing an Open Source Option for NASA Software," NAS Technical Report NAS-03-nnn, https://ntrs.nasa.gov/archive/nasa/casi.ntrs.nasa.gov/20030054432.pdf.

[5] See https://ti.arc.nasa.gov/opensource/nosa/.

[6] C.A. Mattmann, D.J. Crichton, A.F. Hart, S.C. Kelly, C.E. Goodale, P. Ramirez, J.S. Hughes, R.R. Downs, and F. Lindsay, 2012, Understanding open source software at NASA, *IT Professional* 14(2):29-35, doi:10.1109/MITP.2011.118.

[7] See https://obamawhitehouse.archives.gov/sites/default/files/microsites/ostp/ostp_public_access_memo_2013.pdf.

[8] See https://www.nasa.gov/sites/default/files/atoms/files/206985_2015_nasa_plan-for-web.pdf.

[9] See https://sourcecode.cio.gov.

[10] See https://code.nasa.gov/NASA-M-16-21-OCIO-Memo.pdf.

NASA Space Science." The Committee on Best Practices for a Future Open Code Policy for NASA Space Science was formed with the following statement of task:

> The National Academies of Sciences, Engineering, and Medicine will establish an ad hoc committee to investigate and recommend best practices for NASA as it considers whether to establish an open code and open models policy, complementary to its current open data policy. In carrying out the study the committee will:
>
> 1. Review and describe examples of code/modeling policies developed by research teams and communities in the NASA-supported disciplines of Earth Science and Applications from Space, the Space Sciences, and other research communities, as appropriate;
>
> 2. Develop a set of lessons learned from these established approaches—paying particular attention to issues such as, but not limited to, proprietary, export control, code/model maintenance, and documentation considerations;
>
> 3. Define and describe options for policies on open software and open models for research supported by NASA Science Mission Directorate (SMD) and assess the pros and cons of these options from the perspective of the research community and the interests of NASA; and
>
> 4. Recommend a set of best practices for NASA to consider should SMD decide to adopt an open code/open model policy for research supported by the agency. The committee may also choose to present alternate sets of best practices rather than just one recommended set.

The legal and executive directives above motivate this study, but they also occur within a larger context of an international movement toward greater transparency and openness of research as an accepted means to increase scientific rigor, expand knowledge, increase the pace of science, and benefit society. This trend is emphasized in a major National Academies report, *Open Science by Design: Realizing a Vision for 21st Century Research*, which was issued in July 2018 and stressed the benefits of open science, including rigor and reliability; faster and more inclusive dissemination of knowledge; broader participation in research; and effective use of resources.[11] Open source practices are a key part of these, and indeed, fall under the first three major recommendations of *Open Science by Design*, which are listed below, and echo the findings and recommendations of the committee in the current report:

> Recommendation One
>
> Research institutions should work to create a culture that actively supports Open Science by Design by better rewarding and supporting researchers engaged in open science practices. Research funders should provide explicit and consistent support for practices and approaches that facilitate this shift in culture and incentives.
>
> Recommendation Two
>
> Research institutions and professional societies should train students and other researchers to implement open science practices effectively and should support the development of educational programs that foster Open Science by Design.
>
> Recommendation Three
>
> Research funders and research institutions should develop the policies and procedures to identify the data, code, specimens, and other research products that should be preserved for long-term public availability, and they should provide the resources necessary for the long-term preservation and stewardship of those research products.

While the National Academies report discussed above stresses the benefits of open science, as NASA considers whether to establish an SMD-wide policy on open source software (OSS), complementing its open data policy,

---

[11] National Academies of Sciences, Engineering, and Medicine, *Open Science by Design: Realizing a Vision for 21st Century Research*, The National Academies Press, Washington, DC, 2018, https://doi.org/10.17226/25116, pp. 7-10.

some members of the community may call for more explanation of the benefits of OSS to science. The statement of task for this report was to recommend best practices for an OSS policy, not to evaluate the costs and benefits of an OSS policy on NASA science, but a short introduction is relevant. In 2012, Morin et al. proposed the following: "Requiring that source code be made available upon publication would . . . yield substantial benefits—including improved code quality, reduced errors, increased reproducibility, and greater efficiency through code reuse and sharing."[12] More broadly, Sonnenburg et al.[13] list the following advantages: (1) reproducibility of scientific results and fair comparison of algorithms; (2) uncovering problems; (3) building on existing resources (rather than reimplementing them); (4) access to scientific tools without cease; (5) combination of advances; (6) faster adoption of methods in different disciplines and in industry; and (7) collaborative emergence of standards. The benefits can be ascribed alternatively to the open source licensing model, or to the open source development model. For example, improved code quality and fewer errors stem from the latter, while reproducibility and efficiency via reuse stem from the former. The better code quality of OSS, versus proprietary or closed source, has been shown within industry settings.[14] In science, evidence of OSS benefits is still mostly anecdotal, but strong by way of the counterexamples where errors in software have resulted in retracted papers or erroneous trends in data.[15] The efficiency gains from open source development models and code reuse are illustrated plainly with community library development.[16]

In this context, an OSS policy informed by this study is a logical next step for SMD as it moves toward more openness. Developing such a policy for SMD involves complex considerations, in terms of legal and practical constraints, intellectual property, and different software types and applications. Changes in policy are difficult, and the success of a policy can depend on how it is implemented. Accordingly, the committee assembled "lessons learned" from related policy implementations and the community responses. The committee has set forth a number of policy options and recommendations rather than the requested "best practices" for NASA to consider (Task 4 of the statement of task). The policy options and recommendations highlight important considerations in the implementation of an OSS policy at NASA SMD.

1.2 POLICY GOALS

The purposes of any OSS policies that SMD may develop are to serve the goals of SMD and NASA. Based on NASA's vision, mission, and mandates; guidance to the committee from NASA representatives; and the general interests of science and society, the committee identified seven goals.

1. Enhance and enable innovation and discovery.
2. Increase the visibility, access, and reuse of NASA-funded code.
3. Facilitate scientific reproducibility.
4. Encourage collaboration inside and outside of NASA.

---

[12] A. Morin, J. Urban, and P. Sliz, 2012, A quick guide to software licensing for the scientist-programmer, *PLoS Computational Biology* 8(7):e1002598, https://doi.org/10.1371/journal.pcbi.1002598.

[13] S. Sonnenburg, M.L. Braun, C.S. Ong, S. Bengio, L. Bottou, G. Holmes, Y. LeCun, et al., 2007, The need for open source software in machine learning, *Journal of Machine Learning Research* 8(Oct):2443-2466.

[14] See, for example, Coverity, Inc., 2013, Coverity Scan: 2013 Open Source Report, http://softwareintegrity.coverity.com/rs/appsec/images/2013-Coverity-Scan-Report.pdf.

[15] See, for example, G. Miller, 2006, A scientist's nightmare: software problem leads to five retractions, *Science* 314(5807):1856-1857, https://doi.org/10.1126/science.314.5807.1856; J. Irons and J.Bivens, 2010, "Government Debt and Economic Growth Overreaching Claims of Debt 'Threshold' Suffer from Theoretical and Empirical Flaws"; H. Gee, 1998, Satellite climate record in error, *Nature*, doi:10.1038/news980820-1, https://www.nature.com/news/1998/980820/full/news980820-1.html; D.A.W. Soergel, 2014, Rampant software errors may undermine scientific results, *F1000Research* 3:303, doi:10.12688/f1000research.5930.2; D.C. Ince, L. Hatton, and J. Graham-Cumming, 2012, The case for open computer programs, *Nature* 482(7386):485; and B. Boehm, H.D. Rombach, and M.V. Zelkowitz (eds.), 2005, *Foundations of Empirical Software Engineering: The Legacy of Victor R. Basili*. Springer, New York.

[16] A.M. Price-Whelan, B.M. Sipőcz, H.M. Günther, P.L. Lim, S.M. Crawford, S. Conseil, D.L. Shupe, M.W. Craig, N. Dencheva, A. Ginsburg, and J.T. VanderPlas, 2018. *The Astropy Project: Building an Inclusive, Open-Science Project and Status of the v2.0 Software*, arXiv preprint, https://arxiv.org/abs/1801.02634. See also section 2.2 for discussion of community libraries.

5. Maximize NASA's benefit to society.
6. Respect the security and privacy of citizens.
7. Comply with broader government policies.

These goals helped guide the committee's information-gathering process. They provided context for discussing lessons learned from existing policies and community perspectives, policy options, and implementation strategies. They will also guide efforts that SMD develops to assess the effectiveness of the policies they implement.

Overall, the committee operated on the maxim of "as open as possible; as closed as necessary."

For the purpose of this study, the terms *open code* and *open source software* are used synonymously, as defined in Section 2.1. Open source software has generally become a term of choice in the software development community.

1.3 COMMITTEE PROCESS

Aiming to properly gauge the science community's perspectives, and understand possible consequences of an OSS policy, the committee held three in-person meetings that included presentations from diverse stakeholders. The committee also solicited community white papers and received 44 thoughtful submissions that describe a variety of experiences and were both supportive and concerned about an OSS policy. The committee also received legal guidance from an unpaid consultant.

The report is organized as follows: Chapter 2 provides important definitions, a short explanation of relevant legal issues, and an overview of open source licensing models and development models. Chapter 3 reviews existing policies and the lessons learned from the implementation of those policies (Tasks 1 and 2). Chapter 4 summarizes community perspectives with additional lessons learned (Task 2). Chapter 5 presents policy options and recommendations for implementation for NASA SMD (Tasks 3 and 4). Chapter 6 summarizes the committee's findings and discusses implications for SMD.

The charge to the committee was to evaluate options for a NASA OSS policy, not to argue for or against such a policy. Short discussions on the value of open science and OSS are included in Section 1.1 as background, but in general, the report focuses on the information requested in the statement of task.

2

Background Materials

2.1 DEFINITIONS

The committee collected a list of definitions for terms used in this report that are not necessarily defined, or contextualized, when used. The definitions below are aimed at a broad audience. Some references are provided for those interested in a more thorough treatment.

> **Source code**—Human-readable set of statements written in a programming language that together compose software. Programmers write software in source code, often saved as a text file on a computer. The terms *code* and *source code* are often used interchangeably.
> **Software**—This general term is used for computer programs and applications that provide users some degree of utility or produce a result or service. Software can be distributed in executable form, as source code, or as a service via the Internet.
> **Data**—Recorded information, regardless of form, the media on which it may be recorded, or the method of recording. The term includes, but is not limited to, data of a scientific or technical nature, software and documentation thereof, and data comprising commercial and financial information. See NASA Grant and Cooperative Agreement Handbook, Section 1230.60.[1]
> **Metadata**—A form of data that describes and provides information about other data. It often includes information such as author, date created, and a short description of the data, but can also include more extensive information.
> **Open source software**—Software whose source code is under an open source license, by which the copyright holder grants to anyone the rights to inspect, modify, and distribute the source. Synonymous with *open code*
> **Open source license**—A software license, approved by the Open Source Initiative (OSI) as compliant with the Open Source Definition,[2] granting permissions for anyone to inspect, use, modify, and distribute the software's source code for any purpose. Similar standards may be promulgated by other organizations.
> **Derivative work**—A creative work that is derived from or based upon a preexisting creative work and in which the preexisting work is translated, altered, arranged, or transformed in a manner that requires permission from the copyright owner of the original work.

---

[1] See https://prod.nais.nasa.gov/pub/pub_library/grcover.htm.
[2] See http://opensource.org/docs/osd.

Public domain—A work not protected by copyright under the laws of a particular country is in the public domain in that country. Anyone can use it for any purpose without attribution (falsely claiming to have created the work may constitute fraud, however) without violating that country's copyright laws. Some kinds of works (e.g., written laws of nature, typefaces[3]) and the works of some producers (e.g., U.S. government employees, when the work is used in the United States) are not protected by copyright. See Sections 2.3.2 and 2.4.3 for more discussion of licensing and public domain.

Reproducible research—Research published with all the necessary data, source code, and configurations to run the analysis again, re-creating the results and data products.[4]

Replication—A study arriving at the same scientific findings as another study, collecting new data (possibly with different methods) and completing new analyses.[5]

Repository—A central file location that keeps all source code related to a particular software project, accessible to the developers (and possibly a larger community, or the public) through the Internet. This is different from the definition of *repository* commonly used by the data archival community.

Repository branch (or simply, *branch*)—In a code repository, a branch stores a different version of the source code from the main (or *trunk* or *stable*) version. Software developers use branches to make incremental changes or additions to a code base, without interfering with the original trunk version until changes have been tested.

Version control—A system to automatically manage changing versions of a computer file, especially one that contains source code. In software development, version control preserves a complete history of changes to the source code and enables a developer to roll back to an earlier version if needed.

2.2 CATEGORIES OF SOFTWARE

Similar to data, different types of software will require different policy approaches. These different software types are defined in Table 2.1. To understand the complexity involved and how software may evolve over time, the committee considered three general scenarios.

1. A community exists around an open source software (OSS) project: it includes users, developers, and community leaders. Typical examples are the various libraries or tools developed to extend functionality and capabilities of a programming language (e.g., the numerous libraries, such as AstroPy,[6] SunPy,[7] that extend the analysis capabilities of the Python[8] programming language). Researchers who use these libraries may not cite or acknowledge them in research papers, because they are ubiquitous or considered standard. Their open source community empowers researchers at any level in their career to develop software that they may use to conduct their science, enabling more science to be conducted than if the libraries were not available to the community.

2. A large institution, such as a national laboratory or university, supports development of a substantial modeling framework over many years. The framework implements intricately connected source code that calculates equations and descriptions of processes, such that small changes in how one process is formulated may have far-reaching effects in other areas. The team of software developers has in place a rigorous testing and vetting process for incorporating changes into the publicly available version of the software. Community members can download and use the software, but in order to make "official" changes to it, they must work directly with the software development team.

---

[3] CFR Ch. 37, Sec. 202.1(e).

[4] See also D.L. Donoho, A. Maleki, I. Ur Rahman, M. Shahram, and V. Stodden, 2009, Reproducible research in computational harmonic analysis, *Computing in Science and Engineering* 11:8-18.

[5] See also R.D. Peng, F. Dominici, and S.L. Zeger, 2006, Reproducible epidemiologic research, *American Journal of Epidemiology* 163(9):783-789.

[6] See http://www.astropy.org/.

[7] See http://sunpy.org/.

[8] See https://www.python.org/.

TABLE 2.1 General Software Categories to Consider When Developing Policy

Short Name	Name	Description	Examples
Libraries	Libraries and toolkits	Generic tools implementing well-known algorithms, providing statistical analysis or visualization, and so on, which are incorporated in other software categories.	Numerical Recipes, NumPy, general FFTs, LAPACK, scikit-learn, AstroPy, GDAL
Single-use	Single-use utility software	Software written for use in unique instances, such as making a plot for a paper, or manipulating data in a specific way. This software typically uses libraries and could possibly use other analysis software. It is not of general interest owing to its simplicity or specific utility.	Scripts for plotting, downloading data, scripts for producing figures
Analysis software	Analysis, post-processing, or visualization software	Generalized software (not low-level libraries) used to manipulate measurements or model results to visualize or gain understanding. This software often evolves from single-use utility software and may incorporate libraries.	Stand-alone image processing, topology analysis, vector-field analysis, satellite analysis tools, and so on
M&S software	Modeling and simulation software	Software that either implements solutions to mathematical equations given input data and boundary conditions or infers models[a] from data. Includes first-principles models, data-assimilation tools, empirical models, machine learning, mission planning and engineering tools, among others. They often use libraries.	Atmospheric radiative transfer, stellar evolution, upper ocean turbulence, solar wind predictions, orbit propagation (e.g., OpenGGCM, MESA)
Frameworks	Modeling frameworks	Multicomponent software systems that incorporate a variety of models and couple them together in a complex way. Frameworks can include any software category listed above.	Community Earth System Model (CESM) is a collection of coupled models including atmospheric, oceanographic, sea ice, land surface, and other models
Sensor software	Sensor and instrument data processing software	Software for processing uncalibrated sensor measurements into calibrated sensor data and derived data products. This software type applies calibration coefficients, corrections or algorithms, which may be dependent on forward modeling, simulated observations, equations, and data filtering. It may include modeling and simulation software and libraries.	Software designed for processing Satellite Data (SEADAS), THEMIS data processing (THMPROC)
Infrastructure software	Data management and system software	Software used by data centers and large information technology facilities to provide data services.	Metadata Compliance Checker, APIs, Web apps, Giovanni, McIDAS

[a] Models are equations that represent a physical process. In some communities, the term *model* has been used to describe the particular implementation of a set of equations within software.

3. A close-knit group of researchers, or even an individual researcher, develops and routinely modifies some software over a period of several years. The software may be available to the public for download and use (in executable form, i.e., without the source) or may not be publicly available at all. The developers have invested a large amount of time and resources developing the software. They benefit from their intellectual property and choose to develop and use their software exclusively to retain a competitive advantage and advance their careers. A concern exists that if the software is released, others may take it, make changes, and claim ownership, without involving or crediting the original developers. Additionally, developers see the possibility that improper use of the software could negatively impact their academic reputations.

The scenarios above demonstrate some of the different software development pathways that would be affected by an OSS policy. Additionally, NASA formally defines different software types in Appendix D of NASA Procedural Requirements (NPR 7150.2B) "Software Classifications."[9] This appendix defines eight broad classes that

---

[9] See https://nodis3.gsfc.nasa.gov/displayDir.cfm?Internal_ID=N_PR_7150_002B_&page_name=AppendixD.

range from various types of mission-critical, spacecraft-operations software to general-purpose business and desktop software. Based on guidance from NASA and the policy goals described in Section 1.2, this report considers policy for portions of the following three of the eight NASA software classes:

- Class C: Mission Support Software or Aeronautic Vehicles, or Major Engineering/Research Facility Software
- Class D: Basic Science/Engineering Design and Research and Technology Software
- Class E: Design Concept and Research and Technology Software

Software related to operating a spacecraft is out of scope for this report (Classes A, B, and part of C) because of clear safety and national security concerns with their publication (see Section 2.3). Day-to-day software that supports the routine administrative business of NASA (Classes F, G, and H), such as proposal submission systems and basic office software, is also out of scope. The committee thus focused on "science software," as discussed below, because it is most relevant to the policy goals.

Within the scope of science software, such as the global magnetosphere model depicted in Figure 2.1, it is still pertinent to consider different types. A complete or definitive classification scheme for software categories is impractical, and it may be useful to consider multiple schemes when developing a discipline-specific policy implementation. For the purposes of this report, the committee puts forward one possible classification scheme, detailed in Table 2.1.[10] It is important to note that any of these categories can be considered "community software" when developed by a collective effort following open source best practices.

Other conditions that may affect software policy include the following: the funding source, the parties involved, the development history, and the size and complexity of the software and computational requirements. NASA may fund software development in various ways through research grants, through contracts, or through in-house development at NASA centers. Each of these funding mechanisms may incur different intellectual property rights and legal obligations. Where NASA is a partial funder of software, policy options may be constrained by requirements from other contributors. In some cases, collaborative relationships with other agencies may be in place to develop large modeling frameworks. For example, NASA, the National Science Foundation (NSF), and the Department of Defense (DOD) have jointly funded development of solar and space physics frameworks. Large community software projects may involve contributions from different institutions, agencies, and countries. Some software projects rely on components developed elsewhere, possibly including commercial software. Large legacy software projects, developed over extended periods, often have a complex history. For example, a project could include contributions from many developers, some of whom may have passed away but still retain copyright or who may not wish to apply an OSS license, all rights reserved copyrighted software, or software that has a restrictive license. Opening the source code for these projects could be complicated, expensive, and not significantly advance science unless the software is still broadly used. Similarly, "software as a service" or machine-learning software may require different policy implementation (see Section 5.2.4).

Software is not static. Software reuse can increase risk when a package falls into disuse due to lack of maintenance. Single-use software can evolve, mature, and become more broadly useful over time. An analysis tool could evolve to become a library and later become a community-developed and maintained library. When developing policy, it is important to recognize and potentially encourage this evolving maturity of software.

Finding: The range of software types and development scenarios is vast. Policies will need to account for this diversity.

2.3 LEGAL ISSUES

Developing and implementing an open source code policy requires an understanding of the legal framework within which open source code is created and distributed. This framework includes the following:

---

[10] Policy options and how they affect different categories of software are discussed in Section 5.4 and Table 5.1.

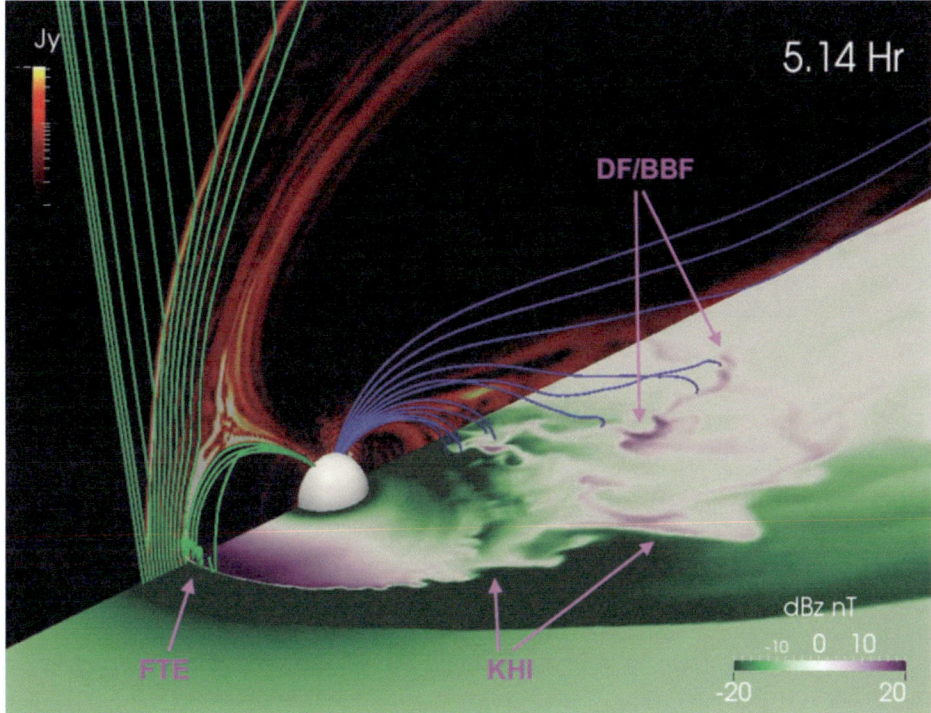

FIGURE 2.1 High-resolution global simulation of mesoscale processes in Earth's magnetosphere as an example of modeling and simulation software. The image shows dipolarization fronts and bursty bulk flows (DF/BBF), Kelvin-Helmholtz instability (KHI), and flux-transfer events (FTEs). SOURCE: M. Wiltberger, NCAR/HAO; V.G. Merkin, JHU/APL; and J. Lyon, Dartmouth College.

- Copyright law and ownership,
- The public domain,
- Patent law,
- Export controls,
- Grant and contract terms, and
- Considerations for institutional grantees.

2.3.1 Copyright Law and Ownership

Copyright gives creators the exclusive right to make copies of an original creative work such as computer software. It also prohibits anyone else from using some or all of the work in another new work, among other exclusive rights. Some uses of copyrighted works are allowed without the holder's permission, such as uses that fall under the doctrine of fair use in the United States. Unless an exception applies, however, in the absence of a license, a user[11] of someone else's copyrighted work infringes copyright and may be subject to fines and penalties. Copyright in the United States generally expires 70 years after the creator's death;[12] therefore, there is little if any software not subject to copyright, and this is expected to hold true for the foreseeable future.

Unless otherwise excepted by law, copyright exists from the moment authors express an original idea in tangible form (e.g., the moment they write an original poem on paper or on a computer). An author need not follow

---

[11] A user in this context is someone running or potentially modifying the software.
[12] Important nuance, because copyright of works made for hire have a different term.

any formalities to claim copyright, such as registering or applying a copyright notice on the work. Generally, when a person creates a copyrighted work within the scope of that person's employment, the work is deemed "made for hire" and the employer owns and controls the copyright. When a person creates a work as a contractor, on the other hand, that person usually owns and controls the copyright unless the contract includes terms saying otherwise. Some works have more than one copyright holder—for example, if two or more people create the work or if different people contribute to the work. Community software projects often have multiple copyright owners, unless the contributors have agreed to assign their copyright to a single organization or individual to manage.

Owners of copyrighted works may grant permissions to others for using the work on certain terms under a copyright license. Licenses may be specifically negotiated, but a copyright holder may also use a public license to grant use permissions to the public at large. Anyone can rely on the public license provided they abide by the license terms.

2.3.2 The Public Domain

Generally, the public domain consists of works free of copyright. This is the case if the copyright term expired, the holder of copyright relinquished their copyright before the term expired, or the work was never protected by copyright. For example, facts and laws of nature are not subject to copyright.

Works by the U.S. government (works created by U.S. civil servants in the course of their official duties) are not protected by U.S. copyright law and are thus public domain in the United States.[13] The U.S. government may hold copyright in those same works under the laws of other countries, however, and it can hold copyrights that are transferred to it (e.g., by contractors). Works that are produced jointly between U.S. government employees and others may be protected by copyright, depending on the terms of collaboration. Note that even where a work is in the public domain, the custodian of the work is not obligated to distribute it or otherwise provide access to it without a law or regulation that requires it be made available. Laws applying to the public domain vary from country to country, creating a complex legal landscape in the internet age. Licenses or dedications (like the Creative Commons CC0 dedication) that make clear how a work may be used worldwide, provide clarity in this complicated landscape.

2.3.3 Patent Law

Whereas copyright protects the expression of an original idea, a patent protects original inventions, including original solutions to engineering problems. A patent gives the owner the exclusive right to prohibit others from making, using, or selling the invention, among other things. Unlike copyrights, patents require filing an application and approval by a national patent office. A patent typically lasts 20 years in the United States. Algorithms and scientific laws cannot be patented, but devices implementing them can be. Software is unique because it is protectable by both copyright and patents. Unlike copyright, the U.S. government can obtain and enforce its patents in the United States and elsewhere, if registered.

The Bayh-Dole Act (1980) allows small businesses, nonprofit organizations, and universities under federal research contracts to own patent rights in inventions they may develop using federal research funding. Previously, the government owned such rights. The act aimed to stimulate the economy by motivating research entities to license and commercialize their patents. This act applies to NASA.

2.3.4 International Copyright and Patent Law

Copyright and patent rights are territorial, which means that every country has its own laws that protect (or do not protect) rights in software, including how long any copyright and any patent rights endure. Most open source licenses apply—that is, require compliance with their terms and conditions—only if the licensed work is protected by applicable copyright or patent law. What is under copyright or protected by patent rights in one

---

[13] 17 U.S. Code § 105, https://www.law.cornell.edu/uscode/text/17/105.

country may not be protected in another, and it may or may not require compliance with a public license when the work is reused. As a consequence, a work in the public domain in the United States as a matter of U.S. copyright law, including software created by U.S. government employees within the scope of their duties, may be subject to copyright under the laws of other countries. This means that a software license applied to such a work applies only when another country's laws applies. Further implications for the development of an open software policy for NASA are detailed in Section 3.3.6.

2.3.5 Export Controls

While most software created by NASA scientists may be released without any restrictions if desired, some software cannot be distributed outside the United States because doing so is prohibited by U.S. export laws and regulations. Throughout this report, the term *export controlled* refers to a number of federal laws and regulations that relate to releasing technology (including software) that may impact national security. The list of policies, laws, and regulations below is not exhaustive, but represents the most common regulations that scientists may encounter.

The International Traffic in Arms Regulations (ITAR) prohibit the export of munitions to enemies of the United States.[14] At least two kinds of software are ITAR munitions: strong encryption and nuclear-reaction simulation. Once software has been labeled ITAR-restricted, it is illegal for anyone in possession of a copy of the software to allow its transport out of the United States. Posting ITAR-restricted software online without sufficient protection against unauthorized copying is also illegal. Software may be published under an open source license and still be ITAR-restricted preventing its distribution outside of the United States.

The Federal Acquisition Regulation (FAR),[15] as supplemented by the NASA FAR Supplement[16] (NFS), establishes "uniform policies and procedures" for acquisitions made by NASA, including software created by contractors. These regulations include provisions that can impact the release of that software by contractors, among other things. Additionally, the Office of Foreign Asset Control (OFAC) of the U.S. Treasury Department enforces sanctions imposed by the United States against other countries.[17] Depending on where and to whom software is distributed, an OFAC license may be required. Similarly, the U.S. Department of Commerce issues Export Administration Regulations (EAR) that regulate the export of certain products from the United States.[18] EAR and ITAR have subtle differences in how they interpret what information is publicly accessible and therefore not subject to export controls. For more information regarding guidance on legal issues, see Section 4.2.2.

2.3.6 Grant and Contract Terms

NASA grants include terms giving the government the nonexclusive right to use the results of research without paying royalties, and even to sublicense those rights. However, research organizations retain ownership of the intellectual property. This means that in the absence of specific terms requiring that software be developed and released under open source licensing terms, the decision about releasing software may be left to university technology transfer offices, or similar entities. Some grant programs, however, specify that software must be released under an open source license—for example, the NASA Earth Science Data Systems program.[19]

Contractors may be subject to contract conditions that place high barriers for open source release. NASA FAR Supplement 1852.227-14[20] "Rights in Data: General" (April 2015), paragraph (4)(i), states the following: "The Contractor agrees not to assert claim to copyright, publish or release to others any computer software first produced in the performance of this contract unless the Contracting Officer authorizes through a contract modification." To apply an open source license to software requires the contractor to assert copyright. In the absence of default

---

[14] See https://www.pmddtc.state.gov/regulations_laws/itar.html.
[15] See https://www.acquisition.gov/browsefar.
[16] See https://www.hq.nasa.gov/office/procurement/regs/nfstoc.htm.
[17] See https://www.treasury.gov/about/organizational-structure/offices/Pages/Office-of-Foreign-Assets-Control.aspx.
[18] See https://www.bis.doc.gov/index.php/regulations/export-administration-regulations-ear.
[19] See https://earthdata.nasa.gov/earth-science-data-systems-program/policies/esds-open-source-policy.
[20] See https://www.law.cornell.edu/cfr/text/48/1852.227-14.

contract modifications that can be applied when NASA implements an OSS policy, the FAR supplement language will remain an impediment to policy implementation.

Finding: The community sees current NASA acquisition regulations as a barrier for open source release of software created under contracts to NASA.

2.3.7 Considerations for Institutional Grantees

NASA funds are generally granted to institutions that in turn distribute the funds to investigators and others while handling financial reports and legal compliance themselves. Generally, in the absence of an agreement between the individual and the institution stating otherwise, all intellectual property created by the recipient resulting from the grant is owned by the institution as a work made for hire. The Bayh-Dole Act prevents the institution from charging the U.S. government for using patented inventions for federal purposes. Some institutions, notably universities, have employment agreements giving researchers ownership rights in the intellectual property they create on the condition that commercialization is handled through the university under a profit-sharing agreement.

2.4 LICENSES—SPECTRUM OF OPENNESS

The Copyright Act of 1976 ensures that any original creative work, including computer source code, is automatically protected by copyright once created, except for works created by the federal government that are excluded by 17 U.S. Code Section 105. When sharing software with the public on a code-sharing platform such as GitHub,[21] copyright still applies and will generally restrict the software's use, unless the owner has granted a license to the user.

Finding: Unless public domain applies, source code without a license is considered "all rights reserved." Opening source code means both making it public and attaching an open source license.

2.4.1 Open Licenses

A copyright license grants others the right to do something with a protected work that copyright would otherwise prohibit. A software license contains a set of permissions and conditions allowing use of the software legally, without infringing copyright. Standardized public licenses have become a popular way to share software openly and broadly. The license options are many, but their terms must comply with standards established by one or more organizations in order to be characterized as open. The dominant open standard is the Open Source Definition, as defined by OSI, an organization that promotes and codifies OSS and licenses.[22] This definition requires that a license contain certain minimum terms and conditions to ensure basic freedoms for users in order to be called an open license. These terms include the rights to redistribute the software, access the source code, and make derivative works, as well as the ability to require that unofficial changes be distinguished from the work of the original author.

Beyond the basic terms and conditions necessary to be considered open, licenses contain other terms that give the public still more permissions. The degree of permissiveness—what users can do with the software under the license and not violate copyright—is often described as falling on a spectrum from most permissive to most restrictive (Figure 2.2).

The most permissive licenses place the least conditions on use beyond the minimum basic freedoms. They often restate the permissions in the Open Source Definition plus some marking and attribution requirements. Examples include the Berkeley Software Distribution (BSD) license, the Massachusetts Institute of Technology (MIT) license, and the Apache license.

---

[21] A website that hosts version-controlled repositories and provides structured collaboration tools around those repositories, https://github.com/.
[22] See https://opensource.org/definition.

Least Restrictive					Most Restrictive
>	>	>	>	>	>
Public Domain	Permissive	Weak Copyleft	Strong Copyleft	Custom Licenses	All Rights Reserved
E.g., CC0	E.g., MIT, BSD, Apache	E.g., MPL, LGPL	E.g., GPL	E.g., NOSA	
No restrictions or conditions on reuse as a matter of copyright, but patent rights may apply, unless also relinquished	Non copyleft license giving anyone permission to use, modify, and redistribute	Copyleft, but not all derivative works must be licensed under same terms	All derivative works must be licensed under same terms	Specially developed for a particular organization or negotiated; nonstandard	

FIGURE 2.2 Software licensing options on the spectrum of openness, from most permissive (*left*), to most restrictive (*right*). While the NASA Open Source Agreement is an open license as defined by the Open Source Initiative (OSI), it is categorized as a custom license because unlike any other license identified here, it has been interpreted as requiring that all modifications be the original works of the person making the modifications. This precludes modifiers from including in their modifications open source software written by others, unlike the other licenses in this chart. Note that the chart is not comprehensive. SOURCE: D. Peters, 2018, Software License Spectrum, https://doi.org/10.6084/m9.figshare.7019780.v1.

The most restrictive licenses require that derivative works, if distributed publicly, be released under the same license as the original. This ensures improvements and changes are shared back with the public. The term *copyleft* is often used in this case. These licenses are considered restrictive (while still being open) because they limit the conditions under which the derivative can be distributed. Examples include the GNU[23] General Public License (GPL), the GNU Lesser General Public License (LGPL), and the Mozilla Public License (MPL). Of these, GPL is generally considered "strong" copyleft and the most restrictive because all derivatives must be GPL, while the other two allow some uses of the original within other software without requiring all components be licensed under the same license.

2.4.2 Other Licenses and Compatibility

Other nonstandard public and custom software licenses exist. Some are compliant with the Open Source Definition or other definitions of open but may be incompatible with the most widely used open licenses. Generally, licenses are compatible if software under different licenses can be combined and distributed under terms that meet the requirements of both licenses (see Figure 2.3). Compatibility is important if software components under different licenses will be combined and distributed together. When combining code under different open licenses, the resulting software can usually be distributed and reused if the terms and conditions of the most restrictive of the licenses are respected.

NASA releases some software developed by civil servants under the NASA Open Source Agreement (NOSA 1.3),[24] which was developed to address the software's lack of copyright under 17 U.S. Code Section 105 of the U.S. Copyright Act.[25] As mentioned in Section 2.3.2, this law states that works of the U.S. government (which includes civil servant-developed software) are not entitled to copyright protection under U.S. law. The existing OSI-approved licenses dominantly rely on the existence of an underlying copyright and its infringement to enforce their terms and conditions. This severely, if not completely, limits their utility in connection with U.S. government-created software in countries like the United States where no copyright is recognized for such works. NOSA 1.3 relies on contract law wherever copyright law is unavailable as a means by which to enforce the agreement's terms and conditions.

---

[23] GNU is a recursive acronym for GNU's Not Unix.
[24] See https://opensource.org/licenses/NASA-1.3.
[25] See https://gpo.gov/fdsys/granule/USCODE-2010-title17/USCODE-2010-title17-chap1-sec105.

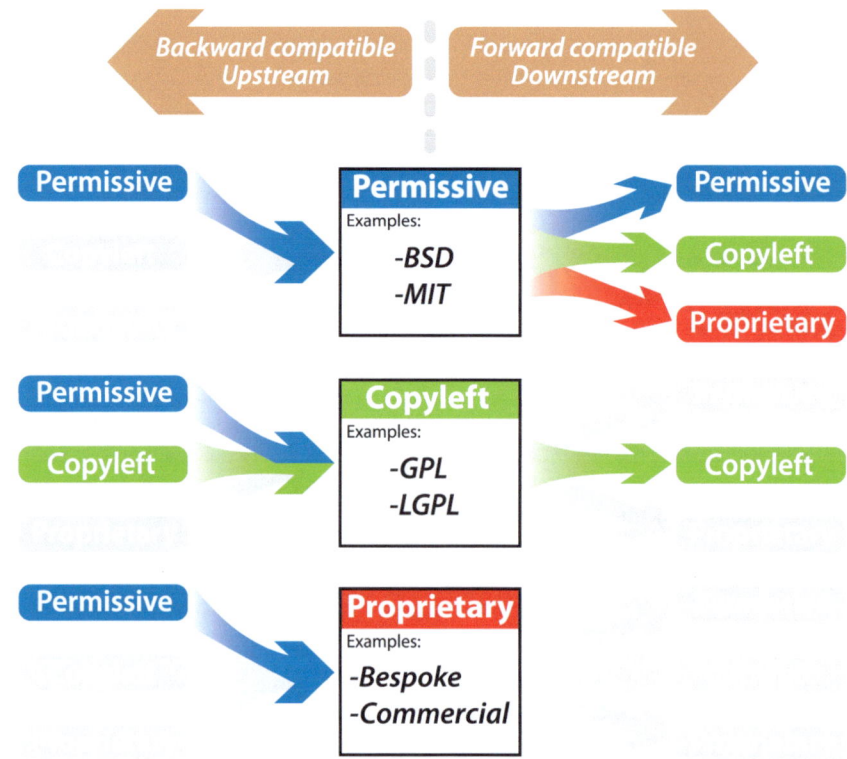

FIGURE 2.3 A schematic representation of license directionality. SOURCE: A. Morin, J. Urban, P. Sliz, 2012, A quick guide to software licensing for the scientist-programmer, *PLoS Computational Biology* 8(7):e1002598, https://doi.org/10.1371/journal.pcbi.1002598.

The OSI agreed that NOSA met the Open Source Definition and approved the license. However, some controversy about specific provisions in the agreement subsequently ensued, and it was determined to be incompatible with GPL. NOSA 1.3 provision 3G states, "Each Contributor represents that its Modification is believed to be Contributor's original creation and does not violate any existing agreements, regulations, statutes or rules, and further that Contributor has sufficient rights to grant the rights conveyed by this Agreement."[26] Provision 3G has been interpreted by some as prohibiting users from including software created by others in their contributions to modified NOSA-licensed software.

This may preclude an individual or organization from modifying the software using portions of code written by others and may limit the utility and potential reuse and improvement, and therefore the impact of software licensed under NOSA 1.3.[27] Others disagree with this conclusion and point to their experience contributing on NOSA projects that include third-party OSS.[28] Provision 3I states, "A Recipient may create a Larger Work by combining Subject Software with separate software not governed by the terms of this agreement and distribute the Larger Work as a single product. In such case, the Recipient must make sure Subject Software, or portions thereof, included in the Larger Work is subject to this Agreement."[29] Hence, there is confusion regarding how this license does or does not allow for software reuse and compatibility with other licenses. There appears to be lack

---

[26] See https://nasa.github.io/CertWare/collateral/2016276-NOSAP3PL.pdf, p. 4.
[27] See also White Paper 28 in Appendix C and Arfon Smith's presentation to the committee.
[28] See http://www.crynwr.com/cgi-bin/ezmlm-cgi/3/17056.
[29] See https://nasa.github.io/CertWare/collateral/2016276-NOSAP3PL.pdf, p. 4.

of mutual understanding between the open source community and NASA patent authorities regarding the interpretation of and need for NOSA 1.3. NOSA 2.0 is an update to NOSA 1.3, developed to be more understandable and to address some of the concerns raised about NOSA 1.3. NOSA 2.0 was submitted to the OSI approximately 5 years ago, but, for unclear reasons, approval is still pending.[30]

Finding: There exists a lack of understanding and clarity within the open source community about the need, desirability, and utility of applying NOSA 1.3 to NASA-funded software, and whether and when other well-accepted and standard OSI-approved licenses may be used instead.

Finding: The community sees NOSA 1.3 Provision 3G as a barrier to contributing to NOSA-licensed software.

Conclusion: If NASA chooses to pursue the development of NOSA, it is important to clarify to the community why NOSA is necessary and why other licenses are inadequate.

2.4.3 Public Domain Versus Licensing for Software

As discussed in Section 2.3.3, software written solely by U.S. civil servants within the scope of their employment is public domain in the United States. Works produced by civil servants in collaboration with nongovernmental employees, on the other hand, may be subject to copyright, depending on the terms of collaboration. Confusion on these points is widespread. The committee found several examples where the U.S. government has asserted copyright over software (e.g., the Common Data Format[31]), but it is unclear whether these works were created in collaboration with nongovernment employees. A common misconception is that software written by federal employees cannot have a license attached: in fact, an open source license is not only allowed but also desirable. It gives permissions globally in those countries where the U.S. government employee's work is protected by copyright. It also signals to other software developers that they are legally permitted to reuse the source code.

The CC0 Public Domain Dedication is a tool that can be used to eliminate any foreign copyrights that apply to a U.S. government work under other countries' laws.[32] CC0 does not license or waive patent rights and is not approved by the OSI. Thus, it is still useful to apply an open source license to software even if it is public domain in one or more countries. This specifies rights regarding patent rights everywhere, and it specifies rights regarding copyrights outside the United States and conveys a preference for how the software is to be used and distributed within the United States. Open source licenses generally help clarify intent and remove ambiguity in the complex global legal framework.

Some federal employees still are convinced that software developed as part of their official duties can only be public domain or released under a CC0 dedication, and some agencies enforce this per policy. The Federal Source Code Policy (Section 7.5), however, clearly recommends the following: "When agencies release custom-developed code as OSS, they shall append appropriate OSS licenses to the source code."[33] Another example is the DOD code.mil initiative. It states: "Even if the code was completely written by U.S. federal employees, it is still good practice to attach a license to the project."[34]

2.5 OPEN SOURCE AS A DEVELOPMENT MODEL

As defined in Section 2.1, OSS refers to software whose source code is under an open source license. From the legal perspective, open source is a licensing model, and this is what enables basic research transparency. However, software reuse requires more than an open license, so any member of an open source community will be quick

---

[30] Per. Comm. Rob Padilla, Chief Patent Counsel/Deputy Chief Counsel, NASA Ames Research Center, and Bryan Geurts, NASA Goddard Chief Patent Counsel.
[31] Common Data Format copyright notice: https://cdf.gsfc.nasa.gov/html/cdf_copyright.htm.
[32] See https://creativecommons.org/publicdomain/zero/1.0/legalcode.
[33] See https://sourcecode.cio.gov/Implementation/#licensing.
[34] See https://www.code.mil/how-to-open-source.html.

to note that open source can also be a development model. The OSS licensing model would do little to enhance reuse without good-quality software in the first place. The OSS development model is behind the creation of that good-quality software.

The essay "The Cathedral and the Bazaar,"[35] by Eric Raymond, is the classic portrayal of the open development model and its capacity for value creation. Its hero, Linux, is the paragon of world-class software developed fully in the open. While the "cathedral" metaphor refers to a closed team writing software and making sporadic releases, the "bazaar" pictures the organized ruckus of true open source development. A simple precondition of having a "plausible promise" of a solution to a problem, one that could be crude and incomplete, is enough to release software under this model ("release early, release often"). By cultivating users as co-developers and taking advantage of the Internet for distributed collaboration, the quality of the software rises quickly. Expressed as *Linus's law* (for Linus Torvalds): "Given enough eyeballs, all bugs are shallow." Experience shows that bug reports are more useful if the user is source-aware: the developers and the users (or beta testers) have a common language and combine their efforts in identifying the issue and proposing a solution. Raymond declares the triumph of OSS: "the closed-source world cannot win an evolutionary arms race with open-source communities that can put orders of magnitude more skilled time into a problem." Box 2.1 details the evolution of NASA's open source Nebula software project into a collaboration with Rackspace, Inc., which then grew into a globally adopted technology resulting in significant financial benefits and job creation.

Open source projects nurture a community, where users are invited to play an active part. *This does not mean that anyone can make changes to the official version of the source code*—a typical misconception. Users or testers can become co-developers by submitting code changes to be reviewed and approved by a merit-based group: the maintainers and the core developers. To facilitate review, core developers build tests and use technology to continuously apply those tests to any submitted code. Open source projects also have governance processes, and appoint leaders that oversee decision making, steward the project, and promote trust. Transparent peer review is at the center of quality assurance and trust building. Ideally, the open development model is highly collaborative, fully transparent, merit-based, and democratic.

Research software—that is, the software that researchers develop to aid their science—often remains of interest to only a small group of people. Such specialized software may not reach a critical mass of users that gels into a community. Even when just a handful of people may use the software, however, the practices that help larger projects manage cooperative development still offer benefits. Technologies and methods such as testing, version control, and code review improve the quality of software irrespective of its scale. In some cases, research software or libraries that begin as tools for a special science workflow do become useful beyond the creators. Over time, they may mobilize new users to become contributors, the need for governance arises, and the project becomes community software.

Finding: The open source development model involves a community of users, but code contributions are sanctioned via peer review and approved by maintainers or core developers. Software can be openly licensed yet not follow the open development model—in this case, it is not considered community software.

---

[35] E.S. Raymond, 1999, *The Cathedral and the Bazaar: Musings on Linux and Open Source by an Accidental Revolutionary*, O'Reilly Media.

> **BOX 2.1**
> **Open Source Software Development Example**
>
> A NASA software project called Nebula, originating at Ames Research Center, spurred innumerable innovations, thanks to its adoption of the open source model. It was born around 2008 from a need to provide a standard set of tools for developing NASA websites, and it grew into a cloud platform solution—long before the information technology world had adopted cloud solutions at scale. The Nebula team recognized that "open source development would facilitate a collaborative environment without borders."[1] As soon as the open source code of Nebula was announced, industry became interested. The company Rackspace, Inc., running one of the largest public clouds in the world, had in-house software to perform similar functions, but recognized that Nebula had features it lacked. Rackspace decided to collaborate in NASA's open source project, and both teams soon saw gains from the collaboration. Rackspace open sourced its complementary software components, and the project became known as OpenStack in July 2010. Since then, more than 100 companies and thousands of developers have joined the project. More than 1,000 participants from around the world now attend the OpenStack conferences. Estimates of the revenue generated by the OpenStack technology are in the hundreds of millions of dollars, with thousands of jobs being created in just a few years. In 2012, the OpenStack Foundation was launched as the new, independent nonprofit home for the OpenStack project. It now has more than 82,000 members from 187 countries.[2] The OpenStack Foundation recently expanded its cloud infrastructure projects to include new technologies such as containerization and continuous integration. A May 2018 white paper[3] reviews the integration of containers with OpenStack, and reports widespread adoption in telecommunications, large-scale research (particularly at the European Organization for Nuclear Research, CERN), cloud storage solutions, and many more high-impact infrastructure technologies.
>
> ---
> [1] See https://spinoff.nasa.gov/Spinoff2012/it_2.html.
> [2] See https://www.openstack.org/foundation/.
> [3] See https://www.openstack.org/containers/whitepaper.

3

Past and Current Policies

The committee reviewed existing policies for data and software, and many of the lessons learned can be summarized by the statement, "Software is data, but data are not software." Software is included in the definition of data (Section 2.1), but software is copyrightable, whereas data are not. The ability to claim copyright is an important distinction that changes how software policies can be implemented versus data policies. As NASA began implementing an open data policy in the 1990s, there was a simultaneous expansion in the volume of data. Initial data policies were limited to the simplistic requirement of a data management plan. As scientists and the data archive centers gained experience, data management plans expanded requirements based on lessons learned from previous data sets (e.g., formal archival centers are a better long-term solution). As NASA open data gained broad acceptance and began to be integrated into unexpected applications, data management plans began to require specific file formats and metadata attributes to ensure consistency across NASA's diverse data sets. Simultaneously, an appreciation of the need for scientific results to be reproducible began gaining momentum within the national and international communities, and open data policies facilitated this need. The culture around NASA space science data has shifted from keeping data closed to open sharing of data. In many disciplines, not publicly sharing data is now seen as antithetical to the goals of science. This transformation of cultural "norms" in the science community is now beginning with software. It could occur more quickly and gain acceptance more broadly if policy is implemented carefully by reviewing lessons learned from existing policies for both data and software. In this chapter, the committee reviews examples of existing data policies, data management plans, and software policies and management plans for NASA Science Mission Directorate (SMD) divisions, other federal agencies, and publishers. Last, scientific journal policies are reviewed as they impact the community through requirements for publication. Only illustrative examples that yielded important lessons to consider are discussed.

3.1 DATA POLICIES

3.1.1 NASA

As discussed in Section 1.1, before the late 1990s, data sharing was cumbersome, involving mailed magnetic tapes, compact disks, or hard drives. The scientist who physically held the data controlled access, thereby limiting scientific advancement and reproducibility of results. Back then, restricting data access, usually to within the science team or individual's research group, was the accepted practice. With the advent of inexpensive digital storage and fast transfer of information over the Internet, it became easier to share data. In 1994, NASA's Earth Science Division (ESD) adopted a full and open data policy, with no period of exclusive access beyond initial instrument calibration periods (usually 6 months), and nondiscriminatory access, for all NASA-generated standard products. This initially applied to Earth Science missions but was adopted by the other divisions. Despite initial resistance to publicly releasing data, open data access with few, if any, restrictions became the new accepted practice.

NASA SMD has been a leader in open data and creating the infrastructures necessary for managing, curating, and disseminating the data from its science missions and programs as well as archiving and providing universal access to science data products. NASA describes its activities as archiving

> All science mission data products to ensure long-term usability and to promote wide-spread usage by scientists, educators, decision-makers, and the general public . . . to facilitate the on-going scientific discovery process and inspire the public through the body of knowledge captured in these public archives. The archives are primarily organized by science discipline or theme. Communities of practice within these disciplines and themes are actively engaged in the planning and development of archival capabilities to ensure responsiveness and timely delivery of data to the public from the science missions.[1]

NASA SMD has had numerous programs that fund mission teams and other experts to document and archive mission data and derived products.

One of the most comprehensive data systems is ESD's Earth Observing System Data and Information System (EOSDIS), which is a core capability in NASA's Earth Science Data Systems (ESDS) Program. "It provides end-to-end capabilities for managing NASA's Earth Science data from various sources—satellites, aircraft, field measurements, and various other programs."[2] EOSDIS processes, archives, and distributes data for "studying the Earth system from space and improving prediction of Earth system change. EOSDIS consists of a set of processing facilities and data centers distributed across the United States that serve hundreds of thousands of users around the world."[3]

In 2016, NASA updated agency policy for data in NASA Policy Directive (NPD) 2230.1 as follows:

> (1) Ensure public access to the results of federally funded scientific research;
>
> (2) Affirm NASA's commitment to public access to information and data arising from technology development programs and projects;[4]

An implementation example of this policy is given by the 2016 Heliophysics Small Explorer (SMEX) Announcement of Opportunity (AO):

> Mission data will be made fully available to the public by the investigator team through a NASA-approved data archive (e.g., the Planetary Data System, Atmospheric Data Center, High Energy Astrophysics Science Archive Research Center, Solar Data Analysis Center, Space Physics Data Facility, etc.), in usable form, in the minimum time necessary but, barring exceptional circumstances, within six months following its collection.[5]

NASA's open data policy has led to increased access to public investments in research and driven investments within NASA to develop infrastructure, such as formal data archive centers. Each division within SMD has created its own data center organization that responds to community specific needs. These communities are actively engaged in the planning and development of archival capabilities. The facilities ensure responsiveness and timely delivery of data to the public, provide data archiving, and provide visualization tools that it would be inefficient for individual researchers to create. Infrastructure such as these archival centers advances the public use of NASA

---

[1] NASA Open Government Plan, 2010, p. 75, https://www.nasa.gov/pdf/440945main_NASA%20Open%20Government%20Plan.pdf.
[2] See https://earthdata.nasa.gov/about.
[3] NASA Open Government Plan, 2010, p. 76, https://www.nasa.gov/pdf/440945main_NASA%20Open%20Government%20Plan.pdf.
[4] See https://nodis3.gsfc.nasa.gov/displayDir.cfm?t=NPD&c=2230&s=1.
[5] See https://nspires.nasaprs.com/external/solicitations/summary.do?method=init&solId={A0C496AC-9B9D-8F7D-A506-B1695BF9BDE8}&path=closedPast.

data and provides expert guidance for users. The NASA data archive centers facilitate finding and using data for applications and research, resulting in increased use of NASA data.

Lessons Learned: Changes in agency data policies prompted changes in accepted practices regarding sharing of data.

NASA's investments in infrastructure allowed NASA SMD to realize the benefits from an open data policy by providing a robust and comprehensive data system for the scientific research community, policy makers, and the public to have consistent access to curated data. This simple but powerful mechanism to access and use the data requires a large and sustaining investment in this data infrastructure.

3.1.2 USGS

In 2008, the U.S. Geological Survey (USGS) changed policy to provide free and open access to Landsat data, resulting in more than 42 million scenes being downloaded around the globe (Figure 3.1). "After the policy change, the average number of scenes obtained from all sources annually per user more than doubled."[6] The National Research Council report *Landsat and Beyond: Sustaining and Enhancing the Nation's Land Imaging Program* summarized the economic value of Landsat data applications:

> Landsat images make critical contributions to the U.S. economy, environment, and security. Specific economic analyses of some of the benefits derived from the Landsat series of satellites demonstrate its great value for the nation. Most of the analyses use imagery provided without charge by USGS, so their value is not set by market forces. However, analyses of just 10 selected applications—including consumptive water use, mapping of agriculture and flood mitigation, and change detection among them—show more than $1.7 billion in annual value for focused operational management in the United States.[7]

Other studies have also demonstrated the general value of open data. A 2013 market assessment estimated U.K. "public sector information" to have an aggregate value to society of between £6.2 billion and £7.2 billion in 2011/12.[8] A 2016 Australian study estimates that "open government data" has potential to generate up to $25 billion per year, or 1.5 percent of Australia's GDP.[9]

Lessons Learned: Open access policies can dramatically increase the economic value and exploitation of federally funded resources and have unanticipated applications that benefit society.

3.2 DATA MANAGEMENT PLANS

As more data was released to the public through open data policies, in the mid-2000 some federal agencies began requiring that proposals for research funding include a data management plan (DMP). A 2013 Office of Science and Technology Policy (OSTP) memo directed the heads of executive departments and agencies to increase access to the results of federally funded scientific research, including peer-reviewed publications and data.[10] The memo called for agencies to

---

[6] USGS, 2013, Users, Uses, and Value of Landsat Satellite Imagery—Results from the 2012 Survey of Users, https://pubs.usgs.gov/of/2013/1269/pdf/of2013-1269.pdf.

[7] National Research Council, 2013, *Landsat and Beyond: Sustaining and Enhancing the Nation's Land Imaging Program*, The National Academies Press, Washington, DC, https://doi.org/10.17226/18420.

[8] Deloitte, 2013, Market assessment of public sector information, Deloitte (UK) for Department for Business Innovation and Skills, http://www.gov.uk/government/uploads/system/uploads/attachment_data/file/198905/bis-13-743- market-assessment-of-public-sector-information.pdf.

[9] See https://www.communications.gov.au/publications/open-government-data-and-why-it-matters.

[10] See https://obamawhitehouse.archives.gov/sites/default/files/microsites/ostp/ostp_public_access_memo_2013.pdf.

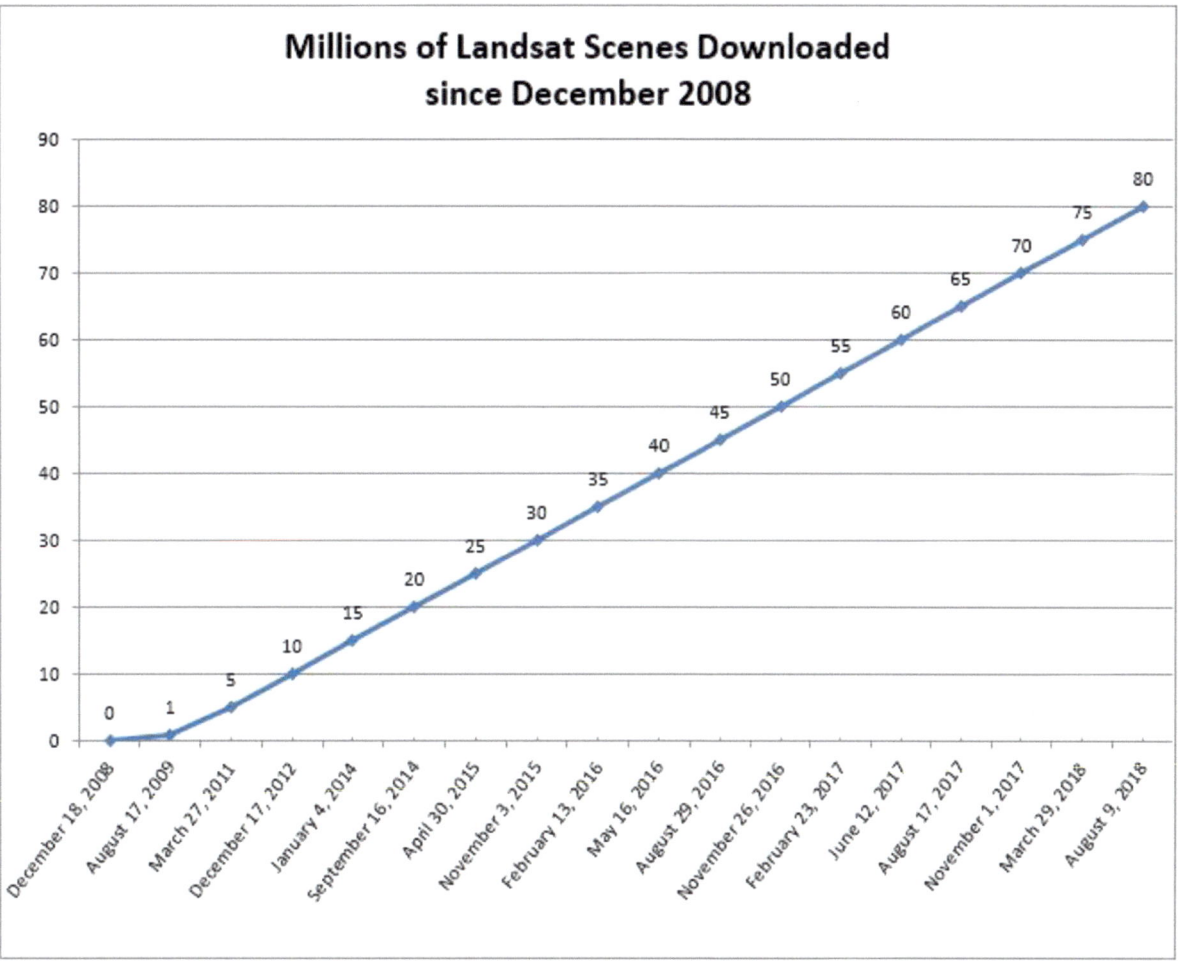

FIGURE 3.1 The Landsat archive became available to all users at no charge in December 2008. As of December 31, 2017, more than 71 million scenes have been downloaded by users worldwide. SOURCE: USGS, https://landsat.usgs.gov/total-landsat-distribution.

Ensure that all extramural researchers receiving Federal grants and contracts for scientific research and intramural researchers develop data management plans, as appropriate, describing how they will provide for long-term preservation of, and access to, scientific data in digital formats resulting from federally funded research, or explaining why long-term preservation and access cannot be justified.

The following sections describe some of the DMP requirements of various agencies and programs. While most DMP requirements concern scientific data, some do mention software, and their handling of this is discussed below.

3.2.1 NASA

The 2009 Earth Venture-1 spacecraft mission announcement was the first NASA funding announcement to call for a DMP: it asked proposed missions to give a "schedule-based end-to-end data management plan, including approaches for data retrieval, validation, preliminary analysis, public release and archiving."[11] Then, research

---

[11] NASA Research Opportunities in Space and Earth Sciences—2009.

funding programs began asking for a DMP as well. Next, a 2010 Mars science program asked that responders follow the Mars Exploration Program Data Management Plan, which is a well implemented, detailed document that describes formatting, access, and archiving.[12] Two Earth Science announcements in 2011 required a simple DMP (to be included in the technical proposal page limit and used as an evaluation criterion), which introduced the need for metadata and data formats.[13,14] In 2012, three Earth Science and one Mars announcement all included required DMPs.[15] One of the Earth Science DMP requirements was expanded from a previous version to include more details on data quality. Individual program managers began realizing the importance of a data management plan before the SMD did, and they began asking for DMPs. The initial requirements evolved, due to community feedback, as both parties' (the program managers and the science community) understanding of data matured.

In response to the 2013 OSTP memo, NASA developed a *Plan for Increasing Access to the Results of Scientific Research*,[16] which addresses only data and peer-reviewed publications. This plan required that all NASA proposals have DMPs that describe sharing and preservation plans for all data used as part of published findings produced during the project. In 2016, the NASA guidebook for proposers required a DMP for all submitted proposals.[17] While the guidebook gave a description of the DMP, individual funding announcements could also ask for additional information and provide evaluation criteria.[18] After 7 years, the DMP was included in the guidebook in a standardized format, integrated into the proposal submission system, and required for all submitted proposals.

Lessons Learned: Subject to discipline-specific needs, providing standardized data management plan submission and formatting as well as educational resources such as the NASA guidebook has facilitated community understanding of new requirements for open data.

The NASA guideline for proposers' DMPs specifically relates to "data generated through the course of the proposed research" and makes passing mention of software. The intended scope of the DMP is to apply to "not only the recorded technical information, but also metadata (describing the data), descriptions of the software required to read and use the data, associated software documentation, and associated data (e.g., calibrations)."[19] The DMP guidelines do not address sharing software, but mention the need to describe software that reads or writes data. At this time, management plans for software are not part of the proposal evaluation except for a few specific funding announcements (see Section 3.3.1).

3.2.2 NSF

The National Science Foundation (NSF) updated its data-sharing policy in 2011 and began requiring an additional maximum two-page DMP (to be evaluated during the proposal review) with basic information about data created during the project.[20] Each directorate may provide additional information and requirements relevant for its community.

For example, the Directorate for Biological Sciences (BIO) routinely updates its DMP requirements. In 2011, BIO required reporting of DMP, which would be evaluated by program managers and committees of visitors, and that future proposals would be evaluated on previous data management efforts. This was an early example of independent evaluation of data-management practice affecting future funding decisions. The evaluation by an independent group is critical to the perception that the new policy will be implemented fairly and evenly. In the 2018

---

[12] NASA Research Opportunities in Space and Earth Sciences—2010.
[13] NASA Research Opportunities in Space and Earth Sciences—2011.
[14] Including the DMP in the technical page limit section of the proposal discouraged including many details about the DMP in order to maximize descriptions of the research.
[15] NASA Research Opportunities in Space and Earth Sciences—2012.
[16] See https://www.nasa.gov/sites/default/files/atoms/files/206985_2015_nasa_plan-for-web.pdf.
[17] See https://www.hq.nasa.gov/office/procurement/nraguidebook/proposer2018.pdf.
[18] See https://www.nasa.gov/open/researchaccess/data-mgmt.
[19] See https://www.nasa.gov/open/researchaccess/data-mgmt.
[20] See https://www.nsf.gov/pubs/policydocs/pappg17_1/pappg_2.jsp#IIC2j.

> **BOX 3.1**
> **Building an Effective Data Policy over Time**
>
> The Section for Arctic Sciences,[1] within the National Science Foundation (NSF) Office of Polar Programs has developed and implemented a rigorous data policy[2] that evolved over many years. The story of that evolution illustrates how policy and compliance can evolve over time and how they require the active participation of program directors along with supporting infrastructure and services.
>
> In the 1990s, one program in the section, the Arctic System Sciences (ARCSS), began to require that researchers funded under this program make their data available at the end of a project. The program also funded a data system to host the data. Researchers were not required to deposit their data in this system if there was another suitable repository, but they were required to submit a metadata record pointing to where the data were archived.
>
> Roughly 10 years later, another program, the Arctic Observing Network (AON), funded a data system and began to require *annual* data deposit. Later, this data system was extended, and all the Arctic programs began to require deposit into the data system.
>
> Program officers monitored compliance with this data access requirement and generally would not approve annual reports for AON, or final reports for other programs, until data were submitted. (Note that investigators cannot receive funding from NSF if they have any outstanding reports due.) Researchers were compliant with the policy, but it was generally known that sometimes researchers would deposit only partial or raw data while keeping the more valuable complete, processed, and quality-controlled data for themselves. In other words, they were following the letter but not the spirit of the policy.
>
> Today, there is a consistent data deposit policy across the entire NSF Office of Polar Programs, including the Arctic Section, with some limited exceptions and minor variance across programs. The Arctic Data Center is funded through a renewable cooperative agreement to archive data and to provide curation services. Program officers do not approve final reports until they receive assurance from the data center that the full data have been deposited and that the metadata is complete to a certain standard. The data center provides services to help the researchers format and deposit their data, complete the metadata, and generally comply with the policy. The program officers act as the enforcement mechanism.
>
> As a result, most all data funded by the division are now readily available and well described. Researchers have come to accept data sharing as routine, but it required about 20 years to achieve.
>
> ---
> [1] See https://www.nsf.gov/staff/staff_list.jsp?orgId=284&subDiv=y&org=OPP&from_org=OPP.
> [2] See https://www.nsf.gov/pubs/2016/nsf16055/nsf16055.jsp.

BIO DMP, the reporting requirements were expanded to include evidence of open sharing by providing the data set location and identifiers or accession numbers that could be easily used to evaluate adherence to the policy. Additionally, the 2018 BIO DMP document included a section on resources providing guidance on data-management practices and writing DMPs, making it easier for scientists to confidently provide the requested information.

Lessons Learned: Clear reporting guidelines and evaluation criteria, as well as evaluation by groups independent from the program, increase the confidence of the community that the policy will be fairly implemented.

It was noted in committee discussion that policy implementation is highly variable across divisions and programs (Box 3.1). There are few NSF-specified formats, and only some disciplines have supported data centers (e.g., Arctic Data Center,[21] Geospace Madrigal[22]). Researchers may find it difficult to find specific data sets, and

---
[21] See https://arcticdata.io.
[22] See http://cedar.openmadrigal.org.

once found, they may find it difficult to use the data, due to custom, undocumented formats. Only some programs require submission to a data center.

Lessons Learned: Program managers are key to successful implementation of policy. Central repositories run by funded expert curators enhance data discovery, usability, and compliance with policy.

3.2.3 USGS

USGS began a program in 2010 to develop formal data management plans with three science centers.[23] Based on this work, in 2012 USGS recognized the need to develop a comprehensive DMP template and funded the three centers to develop one, and to establish best practices, data standards, and lay the foundation for USGS-wide integration. This pilot recognized the importance of policy implementation, providing guidance for scientists, and establishing clear program policy, guidance, roles, oversight, and review mechanisms. The study found that science center "buy in" was critical and that the existing policy resulted in confusion among all interviewed.

For all projects beginning after 2016, USGS requires a DMP with clear guidelines.[24] To help with implementation, the USGS webpage describes the DMP, lists frequently asked questions, and gives examples of DMPs. It also links to "dmptool," a free online tutorial and DMP generator.[25] This well-documented and described plan seems to have directly benefited from the earlier study at three science centers.

Lessons Learned: Pilot studies can provide valuable guidance prior to agency-wide implementation of policy.

3.3 SOFTWARE POLICIES

A fundamental difference between data and software is that writing software creates intellectual property that is covered by copyright laws. Moreover, while data files can be structured to contain all the metadata needed to use the data, software can include multiple interdependent files, resulting in management and version control challenges that are not all fully acknowledged or documented in various federal agencies and programs. This section provides some examples of the evolving nature of software development policies and practices. Although open data policies are becoming the standard, in many fields, software policies are not yet normal practice. The fields that are already implementing open source software (OSS) are benefiting from the large open source community of software developers and the tools to enhance group collaborations. NASA scientists can utilize and build on an extensive infrastructure for OSS—for example, GitHub, GitLab, Bitbucket, or others. Yet, as with open data, support of software management specific to NASA's needs is still required and is typically more complex due to many factors, including multiple software types, as described in Section 2.2.

3.3.1 NASA

Currently, NASA SMD lacks an overarching policy addressing software management, but several programs within SMD have created both ad hoc and official policies. In the following sections, a few of these policies are described, along with some of the lessons learned from these policies.

Earth Sciences

In fall 2015, NASA's ESD published an open source policy for the ESDS.[26,27] The policy was developed to "promote the full and open sharing of all data, metadata, products, information, documentation, models, images,

---

[23] See https://www2.usgs.gov/datamanagement/plan/dmplans.php.
[24] See https://www2.usgs.gov/datamanagement/plan/dmplans.php.
[25] See https://dmptool.org/.
[26] See https://sites.nationalacademies.org/cs/groups/ssbsite/documents/webpage/ssb_174603.pdf.
[27] See https://earthdata.nasa.gov/earth-science-data-systems-program/policies/esds-open-source-policy.

and research results—and the source code used to generate, manipulate, and analyze them." It requires that all software developed through NASA ESD awards, including Research Opportunities in Space and Earth Sciences (ROSES), unsolicited proposals, and in-house funded development, be publicly released with a permissive OSS license. The policy requires using a permissive, widely accepted OSS license and developing the software within a public repository from inception of the funded activity. The software may be granted an exception from the OSS policy for a limited number of stated reasons, including patent or intellectual property law, and national security. Education, software maintenance, and documentation are not yet included in the policy. It states, "NASA will evaluate all funded ESDS software development activities for continued compliance with the OSS policy," but does not discuss how such compliance will be determined.

The first time a ROSES funding announcement references this ESDS OSS software policy is in "A.42 Advancing Collaborative Connections for Earth System Science." The announcement requires, in the description of the proposal contents, that the proposer "describe the software development approach and lifecycle" and "provide an open source software development plan, identify an open source software license and state an open source software release milestone."[28] In other parts of the 2017 NASA research announcement, the OSS policy is not mentioned. The guidelines also do not state where in the final proposal the software development plan is placed and whether or not it is contained within the 15-page limit.

Planetary Sciences

NASA's Planetary Science Division (PSD) has incorporated software management in the DMP for most program elements.[29] Software developed under a NASA grant is to be made publicly available (either at NASA's GitHub organizational account[30] or any appropriate repository) along with sufficient documentation for its use. However, this requirement allows for a number of exceptions based on the practicality of making the software public. Exceptions include software that is not straightforward to implement and software that requires excessive effort to make public. The policy clearly states that software maintenance is not required. An example of PSD modeling and simulation software is depicted in Figure 3.2.

The Planetary Data Archiving, Restoration, and Tools (PDART) solicitation also allows for software tool development and validation.[31] Proposals are required to have a plan for dissemination of the tools and to archive the source code at the NASA's Planetary Science GitHub organizational account.[32] This program also accepts development or enhancement of numerical models, with the expectation that they will be made publicly available.

Astrophysics

The NASA Astrophysics Division does not have a general requirement for software developed within its programs to be made open source, and while the text in ROSES 2018 states that most new astrophysics proposals will require a DMP, there is no similar requirement for software.[33] Several solicitations do independently require new software to be open source, however. For example, the solicitation for the Transiting Exoplanet Survey Satellite (TESS) guest investigator (GI) program states that "proposals must clearly describe the plans to make any new software, higher level data products and/or supporting data publicly available. Software developed with TESS GI funds must add value to the TESS science community, be free, and open source."[34]

Despite the lack of an overarching requirement, the Astrophysics Division has previously taken steps to make commonly used astrophysics software available as open source. For example, the High Energy Astrophysics Science Archive Research Center (HEASARC), the primary archive for high-energy astronomy missions, makes

---

[28] ROSES 2017, pp. A.42-1-A.42-5.
[29] ROSES 2018, p. C.1-7.
[30] See https://github.com/nasa.
[31] ROSES 2018, C.7-3.
[32] See https://github.com/NASA-Planetary-Science.
[33] NASA ROSES, p. D.1-1.
[34] NASA ROSES, p. D.11-4.

FIGURE 3.2 Rendered numerical outputs of model simulations of convective cells and faulting in Europa's ocean and ice shell, as an example of modeling and simulation software in planetary science. SOURCE: Courtesy of Samuel Howell, Jet Propulsion Laboratory, California Institute of Technology.

available an extensive library of general and mission-specific software[35] and the Kepler/K2 missions support a suite of data processing software through the Guest Observer Office.[36]

Heliophysics

Similar to the Astrophysics Division, the committee was unable to find a requirement on software or models developed for any of the currently competed NASA Heliophysics programs, but, as stated earlier, ROSES 2018 does require proposals to include a data management plan.

The NASA Living with a Star (LWS) Science program contained an element called "Strategic Capabilities" that pertained to the development, implementation, and delivery of large computer models for the coupled sun-Earth and sun-solar system.[37] These models were sufficiently mature to address a significant and specific need for achieving the LWS Science Objectives. There has not been an LWS Strategic Capabilities competition since 2011, when the program was renamed NASA/NSF Partnership for Collaborative Space Weather Modeling. The 2011 announcement of opportunity states that "the proposed development must integrate the science into one or more deliverables (e.g., models or tools) broadly useful to the larger community and that will be delivered to an appropriate repository or server site within the term of the project." It later states that "all models and software modules . . . must be submitted to an appropriate NSF and/or NASA modeling center, such as the Community Coordinated Modeling Center (CCMC)." Also, "the proposal must include a description of how the resulting model(s) or other deliverables will be validated,

---

[35] See https://heasarc.gsfc.nasa.gov/docs/software.html.
[36] See https://keplerscience.arc.nasa.gov/software.html#k2fov.
[37] See https://lwstrt.gsfc.nasa.gov/strategic-capability.

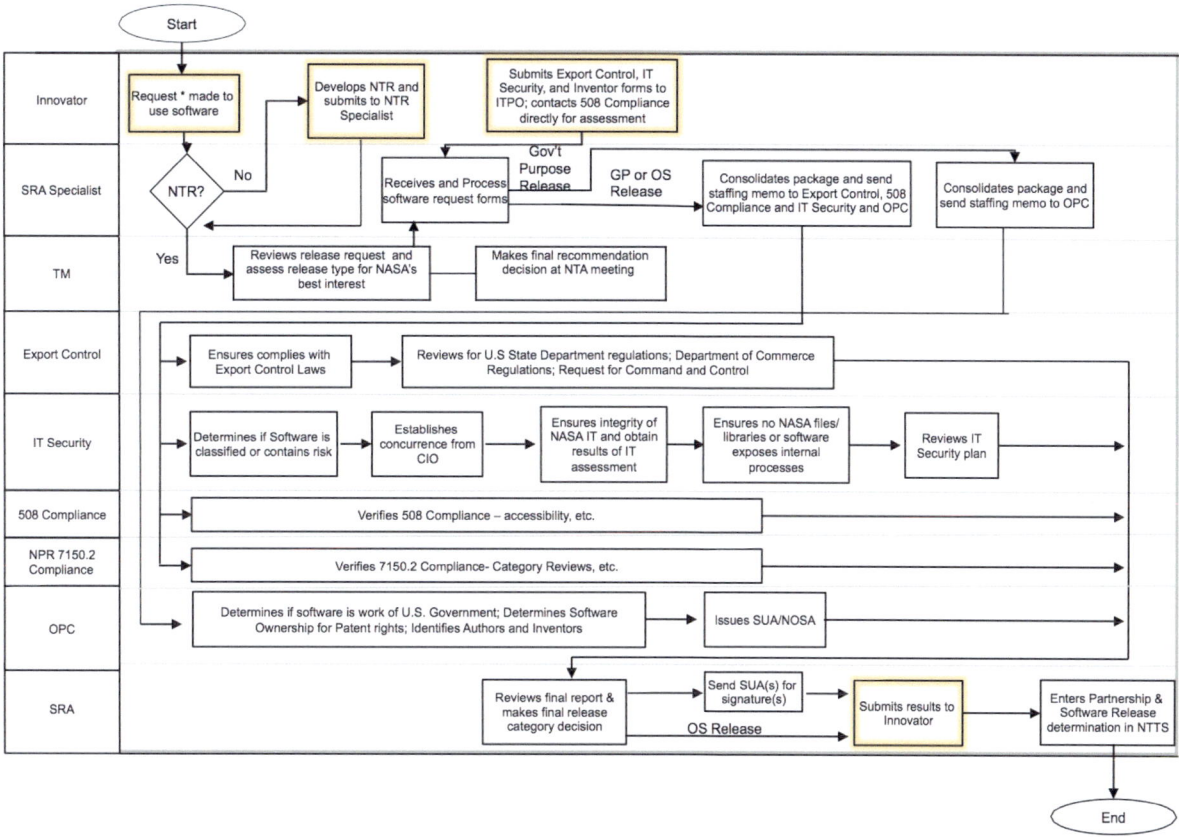

FIGURE 3.3 A flowchart depicting NASA's complex software release process as presented in 2015. NOTE: GP, general public; NTR, New Technology Report; OPC, Office of Patent Counsel; OS, open source; SRA, Software Release Authority; SUA, software user agreement; TM, technology manager. SOURCE: Enidia Santiago, "Innovative Technology Partnerships Office: Software Release Process," presentation to the Technology Education and Assessment Seminar, December 7, 2015.

documented, and made available to potential users."[38] The committee notes that these requirements mention only models and software modules without making specific reference to "open source." Thus, there does not seem to be any requirement to make the source code open. Moreover, since this program has not been competed for 7 years, it is not clear whether the requirements listed in the announcement were enforced or delivered.

Internal Software Development at NASA

While there is no agency-wide requirement to publicly release all internally developed software, NASA employees are encouraged to do so. Such releases are subject to the Technology Transfer Office (TTO) Software Release Process, as described in NASA Procedural Requirements (NPR) 2210.1C and depicted in Figure 3.3.[39,40]

---

[38] See Section 3.3.1. Under the Heliophysics section is page B.7-3 of this announcement of opportunity: https://nspires.nasaprs.com/external/viewrepositorydocument/cmdocumentid=255987/solicitationId=%7B76632D5E-26F6-ACC7-9F82.

[39] See https://nodis3.gsfc.nasa.gov/displayDir.cfm?Internal_ID=N_PR_2210_001C_&page_name=Chapter1.

[40] A partial waiver from these requirements was requested by, and granted to, the Nebula Cloud Computing Platform in 2010 to allow the release of "incomplete" software for open development. More information can be found at https://www.nasa.gov/open/nebula.html and https://nodis3.gsfc.nasa.gov/NRW_Docs/NRW_2210-35.pdf.

At the start of this process, the software developer must first disclose the creation of the software through the filing of a New Technology Report (NTR), in compliance with NPD 2091.1B.[41] The developer may then submit an application to the Software Release Authority (SRA), who will coordinate the review of the software, verifying compliance with export control, information technology security, patent, licensing, and accessibility requirements. Once a release is fully authorized, the software will be made available through an online repository, where it may then be requested by external users. Restrictions on which users will be granted access to the software are dependent on the type of release granted, and most release types will require the signing of a software user agreement (SUA), which can include a nondisclosure obligation or other similar requirements.[42]

All internal software must follow the Software Release Process, as described above, prior to release, regardless of relative complexity or risk, and software that is intended to be fully open to the public is subject to greater scrutiny during the approval process.[43] Because of this, community members have reported a significant increase in the workload and time required to gain approval for the release of certain software through this process, compared to gaining approval to publish the same software in an academic journal (WP 1).[44] In response to such concerns, the TTO has recently developed an electronic document routing system that allows release requests to be issued in parallel and closely tracked, meant to streamline the process and to help identify and correct inefficiencies.[45] While the new tracking system is a clear improvement, this process still does not allow for preapproval of software or expedited approval for simple software of broad utility and could quickly be overcome if the number of requests increased substantially.[46]

3.3.2 NSF

The current NSF Proposal and Award Policies and Procedures Guide states, "Investigators and grantees are encouraged to share software and inventions created under the grant or otherwise make them or their products widely available and usable."[47] Proposals require a DMP that includes a description of software created during the project, but no sharing, licensing, or other details are specified. The comprehensive 2011 NSF Software for Science and Engineering (SSE) Task Force report[48] had 11 major recommendations to support the research, development, and maintenance of OSS infrastructure.[49] SSE was encouraged to develop a multilevel long-term program of support of open source scientific software elements; provide leadership in promoting software verification, validation, sustainability, and reproducibility; develop a consistent policy on OSS that promotes scientific discovery and encourages innovation; support software collaborations among all of its divisions, related federal agencies, and private industry; obtain community input on software priorities and encourage best practices; explore the legal and technical issues with respect to the different open source licenses; promote discussion among its own personnel and with leadership at institutions where its principal investigators are employed; and develop, acquire, and apply metrics for review of SSE projects that are complementary to the standard criterion of intellectual merit. These recommendations have not yet been implemented widely.

While NSF does not have a foundation-wide requirement for open source licensing, the Computer and Information Science and Engineering (CISE) Directorate does include stipulations on a case-by-case basis in various program solicitations. For example, the Future Internet Architecture-Next Phase (FIA-NP) program[50] required all

---

[41] An NTR is required for all new software developed by NASA employees, contractors, and grantees, regardless of whether that software will be released or not; see https://nodis3.gsfc.nasa.gov/displayDir.cfm?Internal_ID=N_PD_2091_001B_&page_name=main.

[42] The five release types are General Public Release, Open Source Release, U.S. and Foreign Release, U.S. Release Only, and U.S. Government Purpose Release.

[43] See https://www.nasa.gov/centers/johnson/techtransfer/technology/software-release.html.

[44] WP [number] is used to reference the white papers submitted to the committee.

[45] D. Lockney, 2018, "NASA Software Release," presentation to the committee on January 18, 2018.

[46] NASA approved 1,369 software releases in 2013 and 5,054 in 2017. Numbers from January 18, 2018, "NASA Software Release" presentation by D. Lockney to the committee.

[47] NSF Proposal and Award Policies and Procedures Guide, p. XI-12.

[48] SSE was formed to identify the needs and opportunities of a scientific open source software infrastructure.

[49] See https://www.nsf.gov/cise/oac/taskforces/TaskForceReport_Software.pdf.

[50] See http://www.nets-fia.net/.

developed software to be released with OSI-approved license.[51] The Secure and Trustworthy Cyberspace (SaTC) program does not require that software be open source, but states that researchers must strongly justify why developed software will not be open source.[52] Any future plans for NSF to require open source on an agency-wide basis were not apparent. These two examples did not specify how to share software or give a specific archive site, provide educational resources on OSS, or give details on how requirements would be enforced. While the 2011 SSE report discussed above presented a clear roadmap for scientific OSS, the existing solicitations do not include its recommendations.

Lessons Learned: The lack of a coherent agency policy at NSF has resulted in inconsistent directorate guidance regarding OSS.

3.3.4 DOE

The Department of Energy (DOE) policies on OSS evolved over time. Release of software at DOE laboratories was allowed in 2002[53] after required approvals from the DOE program and Patent Counsel. In 2003, approval from the Patent Counsel was removed.[54] In 2010, the policy was again changed in response to DOE laboratories' difficulty in obtaining the necessary program approval: affirmative approvals were no longer required except in certain cases,[55] and policy states that DOE programs must be given 2 weeks prior notice to object to licensing as OSS.[56] Laboratories were tasked with monitoring the use of OSS and periodically assessing the value of OSS. While this policy removes DOE approval as a barrier to releasing software, individual laboratories' release policies, which may include overly restrictive export control requirements, may still present barriers (see Section 2.3.5).

Lessons Learned: Removing positive affirmation for software release may reduce agency workloads and streamline software release.

3.3.5 DOD

The Department of Defense (DOD) did a study on OSS with a very different slant than many of the other agencies.[57] It focused on the increasing concern of OSS introducing malware.[58] The report also references some conferences that focus on OSS in the defense arena. This study found that OSS plays a more critical role in the DOD than has generally been recognized. It had the following three recommendations:

1. Create a "Generally Recognized as Safe" OSS list.
2. Develop Generic, Infrastructure, Development, Security, and Research Policies to better develop and utilize OSS.
3. Encourage use of OSS to promote product diversity.

The Army Research Laboratory (ARL) Software Release Process for Unrestricted Public Release provides procedures that ARL government personnel must follow when releasing software source code and software-related

---

[51] See https://www.nsf.gov/pubs/2013/nsf13538/nsf13538.htm.
[52] See https://www.nsf.gov/pubs/2017/nsf17576/nsf17576.htm.
[53] DOE Patent Counsel IPI-II-1-01: Development and Use of Open Source Software.
[54] See https://www.energy.gov/sites/prod/files/2015/01/f19/IPI-OSS%20April%202010.pdf.
[55] Exceptions include export control, software commercialization, or special grant or contract terms and may affect substantial portions of scientific software that are developed in DOE laboratories, https://www.energy.gov/sites/prod/files/2015/01/f19/IPI-OSS%20April%202010.pdf, p. 2.
[56] See https://www.energy.gov/sites/prod/files/2015/01/f19/IPI-OSS%20April%202010.pdf.
[57] MITRE Report Number: MP 02 W0000101.
[58] See http://dodcio.defense.gov/Open-Source-Software-FAQ/.

material to the public, and for accepting software-related contributions from the general public.[59] The document explains why publishing software is important and some of the legal and regulatory constraints on doing so.

Another example of DOD support for open source development is the Defense Advanced Research Projects Agency (DARPA) XDATA program, which develops "an open source software library for big data . . . [including] tools and techniques to process and analyze large sets of imperfect, incomplete data." OSS and peer-reviewed publications are released via the DARPA Open Catalog.[60]

Lessons Learned: Education of software developers and scientists on how to recognize quality OSS and implement it in a secure fashion is essential to maintaining a safe environment.

3.3.6 USGS

USGS has an Instructional Memorandum (IM) establishing the requirements for reviewing, approving, releasing, sharing, and documenting software created by USGS employees and intended for release.[61] The Office of Science Quality and Integrity and Core Science Systems develops and maintains the policy and related procedures. Accordingly, the emphasis is on science software, as follows:

> USGS software intended for public release includes any custom developed code yielding scientific results, thereby facilitating a clear scientific workflow of analysis, scientific integrity, and reproducibility. USGS software releases are made available publicly at no cost and in the public domain. Not all USGS software is suitable for release to the public, including software developed for use on internal Bureau systems or software that has privacy, confidentiality, licensing, security, or other constraints that would restrict release.

The IM requires that released software have an assigned digital object identifier (DOI) and that "software releases and associated documentation constitute official records of the USGS," meaning that they have to be managed according to their records management policies or National Archives and Records Administration (NARA). Section 2.3.4 discusses the issues with public domain licensing. Because there is no OSI-approved public domain choice for releasing software, this policy appears to contradict federal policy discussed in Section 3.3.7.

Lessons Learned: USGS releases software as public domain. In the absence of an OSI-approved option for releasing software as public domain, this choice does not appear to be available to NASA.

3.3.7 Federal Policy

The Federal Source Code Policy's Pilot Program clarified guidance for using open source licenses, as follows:[62]

> Your agency should choose a standard license (or licenses) that can be applied across its open source projects in order to minimize the cost and risk of choosing a license on a project by project basis.
> In choosing your open source license, here are some considerations:
>
> - The Open Source Initiative (OSI) approves open source licenses, a list of which can be found at https://opensource.org/licenses/category. Further still, OSI considers some licenses to be "popular, [and] widely used." Using OSI popular licenses may maximize the interoperability of your open source license with other open source code and increase the comfort level in the minds of potential contributors. OSI maintains a list of popular licenses at https://opensource.org/licenses.
> - Choose licenses that do not place unnecessary restrictions on the code. Any restrictions on the code should be reasonable and essential to furthering your agency's mission.

---

[59] See https://github.com/USArmyResearchLab/ARL-Open-Source-Guidance-and-Instructions.
[60] See https://opencatalog.darpa.mil/XDATA.html.
[61] See https://www2.usgs.gov/usgs-manual/im/IM-OSQI-2016-01.html.
[62] See https://code.gov/#/policy-guide/docs/open-source/licensing.

- Avoid the creation of ad hoc licenses to prevent uncertainty in the minds of contributors as to the legal rights of distribution and reuse. Opt instead to use standardized and well-vetted legal licenses.

As noted in Section 2.3.4, because of how copyright and patent rights differ around the world, a policy that licenses software created by U.S. government employees is not at odds with the fact that as a matter of U.S. copyright law, software developed by U.S. government employees is in the public domain. This is because such software may not be public domain in other countries where the license is useful for sharing works otherwise restricted by copyright.

3.3.8 Large Community Software Projects

Community-Developed Open Source Software Projects

Research computing has historically involved both proprietary and community-based software. A few software tools predominate in certain Earth and space science fields, some of which were developed by a broad community of scientists. Many of these software projects were developed in collaboration with the computer science and computational physics communities with funding from large grants by DOE or NSF, or through centers such as NCAR. These efforts typically have included both extensive verification and validation testing as well as regression test suites. Often, major code releases are tied to journal publications and credit for code developers is facilitated through citations to these papers. For example, a paper describing the SolarSoftWare set of integrated libraries, databases, and utilities has more than 400 citations.[63] In some cases, the community is allowed to modify and contribute to the code and its test suites. But in many cases, code contributions are limited to scientists with specialized expertise. Software developed to solve a specific problem is typically made public after the initial science-application papers are published. Software support in these projects ranges from minimal documentation to fully supported help desks.

Software development in these projects using best practices has been well funded (typically, at $1 million to $3 million per year for 5 to 10 years). Projects that do not have this level of funding usually have less testing and software support. Generally, in a situation where two projects exist addressing the same application, software with less testing and support has fewer users because the software is less trusted in the community. The software projects with large number of users and developers (see Box 3.2) have contributed to many advancements in science that would not otherwise have been possible.

Lessons Learned: Large OSS frameworks provide substantial value to the community, especially when well supported.

Community Coordinated Modeling Center

The Community Coordinated Modeling Center (CCMC)[64] was established in 2000 by a multiagency partnership including NASA Heliophysics and NSF, and it hosts an expanding repository of heliophysics research modeling software and coupled modeling chains (Figure 3.4). The software packages held by CCMC are generally included in the "simulation software" and "modeling framework" categories of Table 2.1, but in this community, they are referred to as "models" and will be referred to as such in this section. Models are provided for use by the scientific community without disseminating source codes. Users of the CCMC perform *runs-on-requests* (RoR), which execute simulations using a Web interface or run simulations with staff assistance. Results of these custom simulation runs are archived and made available for Web-based visualization and analysis or through downloads. The majority of CCMC users are not modelers but use the simulation results in their research rather than in developing software.

---

[63] Number of citations given in Google scholar, https://scholar.google.com/scholar?hl=en&as_sdt=0%2C5&q=Data+analysis+with+the+SolarSoft+system&btnG=.

[64] See http://ccmc.gsfc.nasa.gov.

> **BOX 3.2**
> **An Open Project: Community Earth System Model**
>
> The Community Earth System Model (CESM) was jointly developed and funded by the National Science Foundation (NSF), the Department of Energy (DOE), NASA, and the University Corporation for Atmospheric Research (UCAR) and is depicted in Figure 3.2.1. The National Center for Atmospheric Research (NCAR) administers the CESM, a framework model in Table 2.1, which has been used to conduct experiments in support of the Intergovernmental Panel on Climate Change (IPCC) reports. The framework model has supported releases of the CESM, documentation, discussion groups, a CESM advisory board, and a scientific steering committee. The discussion forums where users and developers can get assistance with CESM also support the individual models within the framework, including: Community Atmosphere Model (CAM), Whole Atmosphere Community Climate Model (WACCM), Chemistry Climate Model (CCM), Community Land Model (CLM), and Community Ice Code (CICE). In addition, forums include support for ice sheet modeling, ocean modeling, biogeochemistry modeling, and paleoclimate modeling.
>
> CESM is freely available through NCAR, but users and developers have to register and agree to the terms of use in order to gain access to the software. The official CESM policy for software changes is as follows:
>
>> All changes to the "official" version of CESM will be brought through a working group with a recommendation to the Scientific Steering Committee (SSC) for approval. Any change must
>>
>> 1. Have results freely available on the Web,
>> 2. Have code freely available on the Web when adopted by the SSC, and
>> 3. Have documentation available as soon as possible.
>>
>> The documentation is the responsibility of the developer.
>
> Support for this public domain project at NCAR requires approximately 50 full-time equivalent people and organization of community meetings and a visitor's program, with funding from a wide variety of sources both internal and external to NCAR and NSF. This program has ensured the open sharing of community framework models that have influenced international policies on climate change. The replicability of results would not be possible without this large effort. The community organization and contributions to CESM have led to substantial improvements in the ability to model climate variability. There are more than 1,500 peer-reviewed journal articles that specifically reference CESM,[1] showing the power of a community-developed and fully supported model.
>
> ---
> [1] University of Washington Library search on April 24, 2018, for "CESM" and peer-reviewed journal articles.

Submission of a model to the CCMC constitutes explicit permission by the developers for public access to model runs and outputs. The CCMC staff is permitted to modify model software only for the following purposes: adapting models to CCMC-specific hardware or converting model input/output formats to CCMC-specific formats. Other types of code modifications require explicit permission from model developers on a case-by-case basis. Permissions to introduce modifications to the source codes on request of RoR users are often granted by developers. All codes at the CCMC are protected from unauthorized access or unintentional dissemination, and by default, source code may not be distributed to any entity.

The benefits to the CCMC model developers became apparent a few years after the start of the center. Modelers noted broader utilization of their models and an increase in the number of publications and presentations citing the model. After the initial slow growth, and some community resistance, when scientists resisted submitting their model to the CCMC, the number of models hosted at the CCMC grew rapidly alongside community acceptance

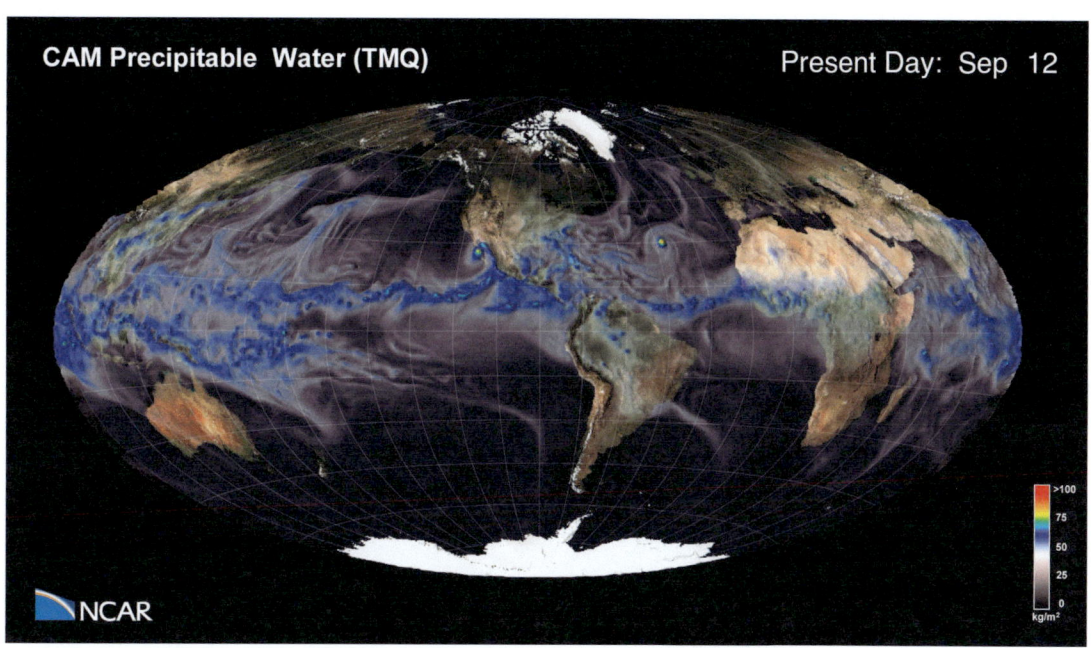

FIGURE 3.2.1 This is a single frame from an atmosphere-only climate simulation of the high-resolution (horizontal 1/4 degree) Community Earth System Model (CESM1). The color scale shows the total water vapor (precipitable water) contained in the atmosphere model. This research used resources of the Argonne Leadership Computing Facility at Argonne National Laboratory, which is supported by the Office of Science of the U.S. Department of Energy under contract DE-AC02-06CH11357 with computing time provided by the Innovative and Novel Computational Impact on Theory and Experiment (INCITE) program. SOURCE: Data simulation—Susan Bates and Nan Rosenbloom, CGD/NCAR; visualization and post production—Tim Scheitlin and Matt Rehme, CISL/NCAR.

(Figure 3.5). While delivering software to the CCMC is now the accepted practice, it took more than 5 years of evolutionary transition to achieve community acceptance of the CCMC and to go from 1 model to 15. It then took another approximately another 5 years before delivery models to the CCMC became a norm and a requirement for some funding opportunities.

Using a modeling center such as the CCMC allows running large-scale models that can take a large amount of computational resources and can have an extremely steep learning curve. Modeling centers bypass this learning curve by having the actual simulation conducted by modeling experts on NASA computing resources, such that the user needs to know only how to interpret the results. While software at the CCMC is not open source, this approach created a mechanism where researchers could share model results without sharing their model. This increased model utilization and reproducibility, but it is not equivalent to an "open source policy."

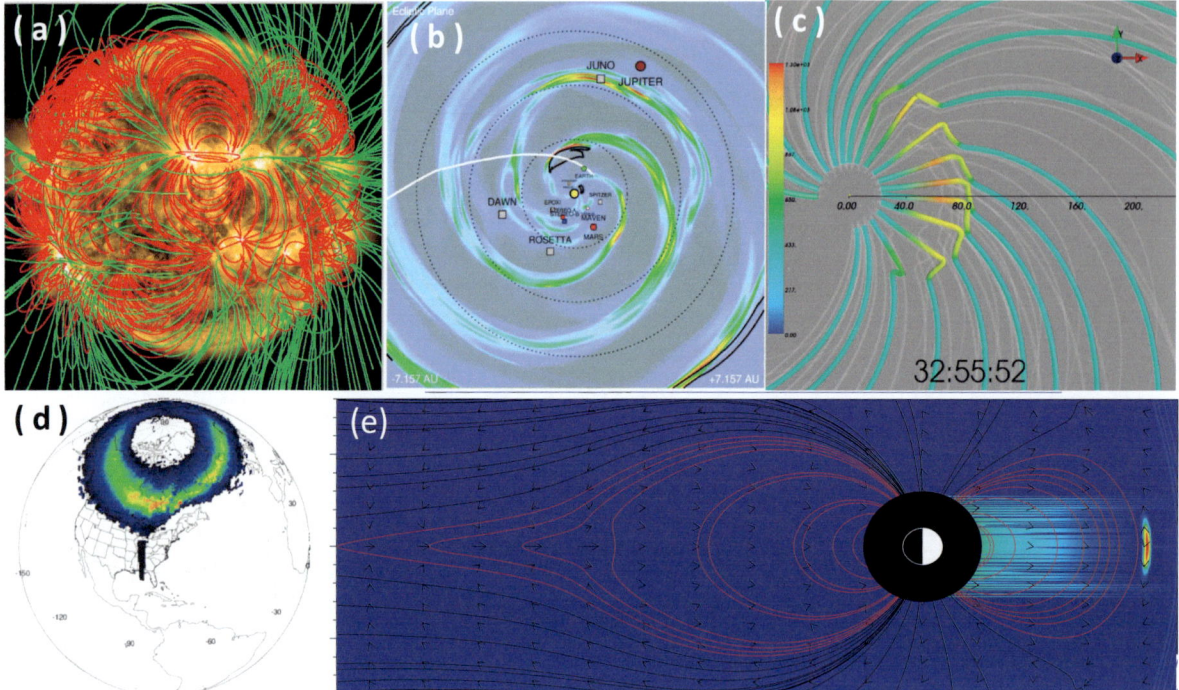

FIGURE 3.4 A selection of visualizations from models hosted at the Community Coordinated Modeling Center (CCMC): (a) nonlinear force-free model of coronal magnetic field; (b) global heliosphere during New Horizons Pluto flyby; (c) energetic particles acceleration in corona and heliosphere; (d) model of aurora; (e) SWMF simulations with a localized region of high resistivity added to the nose of Earth's magnetosphere. SOURCE: Simulation results have been provided by the Community Coordinated Modeling Center at Goddard Space Flight Center through their public Runs on Request system (http://ccmc.gsfc.nasa.gov). (a) The Non-Linear Force Free Coronal Magnetic Field Model was developed by T. Tadesse at NASA and T. Wiegelmann, MaxPlanck Lindau; (b) the ENLIL Model was developed by D. Odstrcil at NOAA and University of Colorado at Boulder; (c) the Corona-Solar Wind Energetic Particle Acceleration (C-SWEPA) Model was developed by N. Schwadron, University of New Hampshire; (d) the Ovation Prime Model was developed by P. Newell, Johns Hopkins University Applied Physics Laboratory; (e) the Block-Adaptive-Tree-Solarwind-Roe-Upwind-Scheme (BATS-R-US) was developed by T. Gombosi, University of Michigan.

Making the models ready for utilization outside the team of original developers requires considerable effort. Many codes are delivered to the CCMC without documentation or comments and require a considerable investment of effort to enable implementation. This has become problematic, because the inflow of models for on-boarding at the CCMC is increasing. Information technology security requirements that include frequent operational system (OS) upgrades also introduce challenges for the long-term usability of models. For example, some models work only on a specific OS and a specific version of a compiler. Any OS upgrade comes with a threat of removing the model from service. To ensure timely implementation of the outcomes of large model development efforts (e.g., NASA living with the Star Strategic Capabilities), the CCMC staff of 13 people (including scientists, software developers, and system engineers) is working with developers on earlier stages of the project.[65] To streamline future upgrades and ensure portability and long-term usability, the CCMC staff help to implement and educate modelers in technologies (such as Docker containers[66]) that are helping to address this problem.

---

[65] See https://ccmc.gsfc.nasa.gov/community/LWS/lws_mod.php.
[66] See https://www.docker.com/what-container.

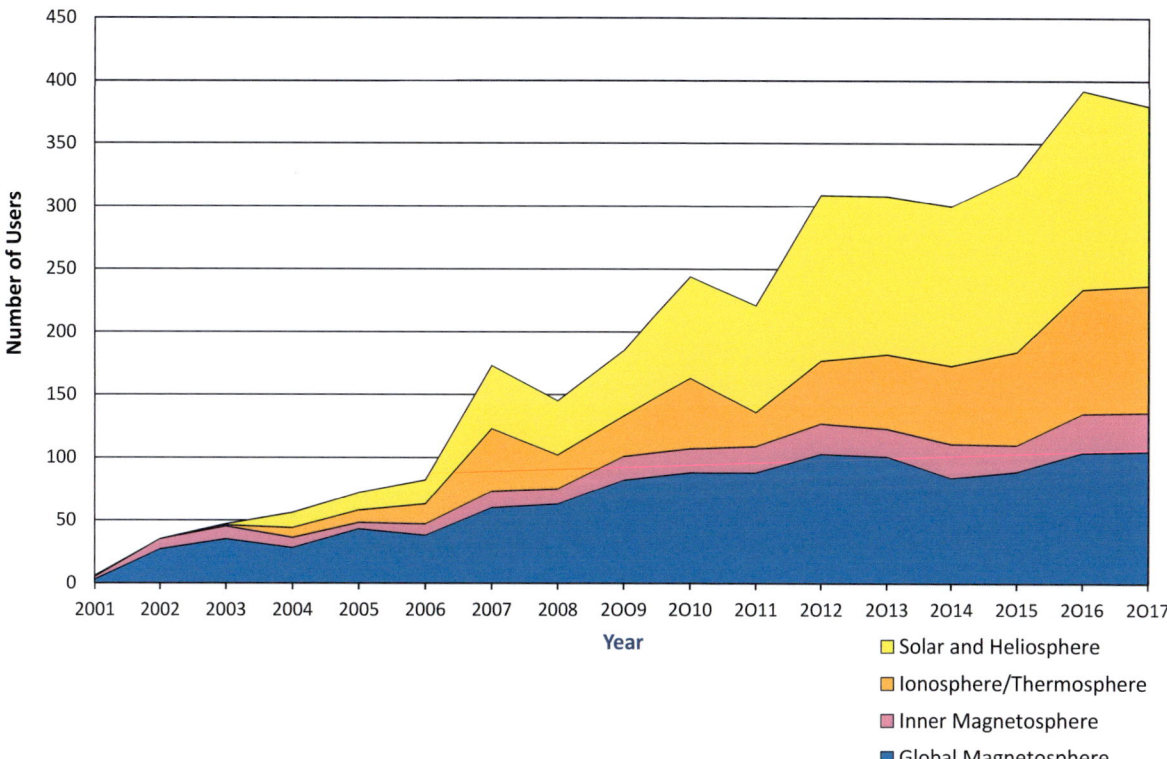

FIGURE 3.5 The number of users of the Community Coordinated Modeling Center's models has increased. In 2017, there were 375 unique users from more than 170 countries. Requests were made for more than 19,000 simulation runs of more than 100 models. Users requested about 8,000 visualizations per month and since 2006 produced more than 300 publications (20 to 50 publications per year). SOURCE: Courtesy of NASA Goddard Space Flight Center.

CCMC users frequently ask to introduce modifications into the source code of a model to achieve their research goals. Some models come with comprehensive documentation and have used best practices for coding and commenting, and the CCMC scientists are readily addressing user requests with model developer permissions. Such models usually have a large team of developers located at multiple institutions.[67] A good example is the Space Weather Modeling Framework[68] developed at the Center for Space Environment Modeling at the University of Michigan, where a source code modification (introduction of a floating resistive spot at the subsolar point of a dayside magnetopause) was introduced in support of the research by Borovsky et al. (2008).

For many other models, introducing modification into source codes can be problematic because most heliophysics models and coupled modeling systems that are utilized through the CCMC runs-on-request system have not been designed for broader community involvement in further development. To find the right place in the source code can be time consuming and sometimes is not achievable without the help of the original model developers. Many models are very sensitive and become unstable after even minor modification. To address such requests, the

---

[67] G. Toth et al., 2005, Space Weather Modeling Framework: A new tool for the space science community, *Journal of Geophsyical Research Space Physics*, 110:A12226, doi:10.1029/2005JA011126.

[68] J.E. Borovsky, M. Hesse, J. Birn, and M.M. Kuznetsova, 2008, What determines the reconnection rate at the dayside magnetosphere?, *Journal of Geophysical Research* 113(A7):210.

CCMC contacts developers of the models for help or advice. This support from developers is not funded, causing some requests to be delayed or declined.

Lessons Learned: The CCMC has been a successful way to increase public access to software but it is not open source, thus it requires additional infrastructure to run the software and control access. This approach to sharing software, which may be appropriate for some communities, is an example of evolutionary transition to new norms and implementation of evolving policies and requirements that led to achieving desirable goals with moderate investments and limited community resistance.

Lessons Learned: Technologies that enable code portability can lessen the concern that legacy models will become unusable.

Lessons Learned: Original model developer involvement may be necessary to introduce modifications into complex legacy codes that were not originally designed for broader community utilization and typically are not well documented.

3.4 JOURNAL POLICIES ON OPEN DATA AND SOFTWARE

The number of journals and publishers with official data policies has been increasing in response to challenges to the validity of published science. Their policies vary widely in what they require from authors.

Wiley is an international scientific, technical, medical, and scholarly publishing company that is also the publishing partner of the American Geophysical Union (AGU). Wiley encourages open data access but allows individual journals to set their own policies.[69] In 1997, the AGU formulated a data policy that encouraged openness.[70] In 2013, this policy was updated to apply to AGU journals

> To advance scientific exploration and discovery, and allow a full assessment of results presented in AGU's journals, all data necessary to understand, evaluate, replicate, and build upon the reported research must be made available and accessible whenever possible. For the purposes of this policy, data include, but are not limited to, the following: Data used to generate, or be displayed in, figures, graphs, plots, videos, animations, or tables in a paper. New protocols or methods used to generate the data in a paper. New code/computer software used to generate results or analyses reported in the paper. Derived data products reported or described in a paper.[71]

This policy is strict in that all data used in a publication and all new code or computer software used in the research must be publicly available. The policy was enacted to ensure reproducibility and transparency in research, and although those reasons were first stated in 1997, it did not become official journal policy until 2013.

The American Meteorological Society (AMS) data policy states that they are

> Committed to promoting full, open, and timely access to the environmental data, associated metadata, and derived data products that underlie scientific findings (see the 2013 AMS policy statement). These data and metadata must be properly cited and readily available to the scientific community and the general public. At initial submission, authors must confirm that the datasets are archived and cited/referenced properly. Likewise, peer review editors are asked to ensure that this AMS expectation is being met.[72]

Reviewers are asked to ensure that data are placed in a location where others can freely download the data. Unless explicit permission is given, it is no longer acceptable to use the language that "the data are available upon request." AMS policy has gradually become stricter with time.

---

[69] See https://authorservices.wiley.com/open-science/open-data/index.html.
[70] See https://sciencepolicy.agu.org/files/2013/07/AGU-Data-Position-Statement_March-2012.pdf.
[71] See https://publications.agu.org/author-resource-center/publication-policies/data-policy/.
[72] See https://www.ametsoc.org/ams/index.cfm/publications/ethical-guidelines-and-ams-policies/data-archiving-and-citation/.

American Astronomical Society (AAS) journals have published papers on software with relevance to research in astronomy and astrophysics.[73]

> Such papers should contain a description of the software, its novel features and its intended use. Such papers need not include research results produced using the software, although including examples of applications can be helpful. There is no minimum length requirement for software papers. If a piece of novel software is important to published research, then it is likely appropriate to describe it in such a paper.

AAS highly recommends using an open source license and a standard software repository and provides guidance to ensuring code citation.

Although the *American Journal of Political Science* (AJPS) is not commonly used to publish articles for NASA-funded projects, it is included here because of its strict open policy. AJPS requires that all data and code used in the publication to be publicly available, specifically so that figures are reproducible exactly as published. Once the review process is completed, an outside company ensures compliance (at a cost of approximately $1,000 per article). This strict policy guarantees reproducibility, and publication in AJPS can be used to verify enforcement of open data and software policies to funding agencies.

Among others, the journal *Astronomy and Computing* recommends that the software developed to produce results in a paper be made accessible.[74] The *Journal of Open Source Software*, on the other hand, was designed to provide a way for open-software developers to obtain citations for their software. The authors submit a short article about their software (required to have an OSI-approved license), and the peer review process checks the quality of the software itself, including functionality, documentation, performance claims, installation instructions and community guidelines.[75]

> **Lessons Learned:** Journals and publishers are moving forward in support of a more open science environment, providing both enhanced recognition and access to data and software. Journals first moved forward with open data requirements that gradually became stricter, then moved to open software requirements.

---

[73] See http://journals.aas.org/policy/software.html.
[74] See https://www.elsevier.com/journals/astronomy-and-computing/2213-1337/guide-for-authors.
[75] D.S. Katz, K.E. Niemeyer, and A.M. Smith, 2018, Publish your software: Introducing the Journal of Open Source Software (JOSS), *Computing in Science & Engineering* 20(3):84-88, doi:10.1109/MCSE.2018.03221930.

4

Lessons Learned from Community Perspectives

The committee solicited community input via a call for white papers issued in December 2017 and presentations on targeted topics at two face-to-face meetings in November 2017 and January 2018. Appendix C includes the text of the white paper solicitation, a list of the white papers received, and a list of presentations made to the committee.[1] The committee strongly values the community input, and all papers were read and considered.

The white papers and the oral presentations articulated numerous areas of impact (both positive experiences and concerns) of a potential new open source software (OSS) policy for NASA's Science Mission Directorate (SMD). The input was aggregated and organized here into big-picture topics, more-focused implications, and procedural areas of impact.

Overall, the community input expressed broad support for OSS. Openness and transparency are seen as central to scientific validity and reproducibility, but various challenges appear in the implementation of policy. A majority expressed positive experiences in opening code, describing a range of advantages, including efficiency, greater collaboration, and more robust code. Many white papers, however, emphasized issues and even pitfalls when trying to regulate the open sourcing of software. Concerns included legal ramifications, institutional barriers, costs, and the impact on individual scientists and their careers. Some suggested that an open source policy may not always benefit science, because for researchers, time spent publishing software comes at the expense of time spent doing science.[2] While an open source policy may enhance science for other researchers, it could be at the expense of the original researcher's scientific output. In addition, there are concerns that researchers may lose motivation to push the boundaries of innovation in their software if they know that they have to immediately release it to the general public instead of having several years to take advantage of the new technology, potentially leading to less innovation in software development. Since doing science and developing OSS are different but complementary activities with different motivations and outcomes, OSS policies may be more successful if they clearly identify value in both activities. Many concerns reflect misunderstandings about open source licensing and processes. Others reveal legitimate legal and institutional barriers. Most unease stems from the culture of how science is currently competed and conducted. While supporting the principle of open dissemination of federally funded research, several white papers emphasized that substantial support is required to strengthen NASA computational capabilities and to build and sustain a successful OSS program.[3] Some also find the comparison between open data and OSS to be mislead-

---

[1] White papers are referred to by number, and titles and authors are listed in Appendix C.
[2] White papers (WPs) 21, 29, 42, and 44.
[3] See, for example, WP 22.

ing and suggest that software is more analogous to instruments.[4] NASA SMD will need to address these concerns as it develops policies, but perhaps more importantly, it needs to foster a new culture of openness and encourage a social norm of sharing and collaboration. Work toward a cultural norm of openness has already begun with open data policies, support, and infrastructure. It needs to continue with carefully constructed support for OSS, beyond simple policy development and implementation. To achieve the full benefits of OSS, it is important to consider how a policy will interact with community norms.

Finding: The NASA science community generally recognizes the value of open source software and supports the principles of openness, but concerns prevail on the details of implementation and the impact on science and scientific careers.

Recommendation: NASA Science Mission Directorate should explicitly recognize the scientific value of open source software and incentivize its development and support, with the goal that open source science software becomes routine scientific practice.

This chapter addresses the perceptions of the community and the current state of OSS and provides some conclusions and recommendations for NASA SMD to remove barriers and foster the move toward openness, regardless of the choice of future policy.

4.1 IMPACT OF OPEN CODE

4.1.1 Software Reuse

Software reuse is a valuable result of open software. Well-written and well-documented community-developed OSS that is reused by many researchers reduces not only labor costs but also the chance of unintended errors (e.g., AstroPy, GDAL, scikit-learn; see WP 24). Instead of writing all analysis software oneself, using a community library can reduce the time spend coding, allowing the focus to remain on designing the analysis and interpreting results. An open code policy can also enhance open data policies, since data that is technically "open" can be effectively locked away for broad-based studies if the source code needed to read the data is not also open (WP 40). Open source can also improve the longevity of software. Well-documented and open software remains available and usable even if the original authors change institutions or retire. This ensures that data from experiments that may have cost millions of dollars in public funds can continue to be accessed and are not lost to future use due to closed software.[5] OSS can protect individual researchers by ensuring that they can always work with their creations even when they change institutions. Additional users may discover new things about their code, including bugs and new unanticipated applications.

Creating high-quality software requires more effort than simple single use software, whether open or closed. Code needs to be well documented and capable of responding gracefully to inadequately specified or incorrect arguments. The costs associated with development of an OSS policy will be discussed further in Section 4.3.

While some white papers, as discussed above, were positive about software reuse, some community members expressed reluctance to making software open source because of concerns about misuse.

1. A user could apply a piece of software in a regime where it is invalid or may not recognize a numerical issue in a simulation result.[6]
2. The reputation of the original software developer might suffer due to errors made later by others.[7]
3. The original developer may have to spend time ensuring that software is used correctly.[8]

---

[4] See, for example, WP 29 and 31.
[5] WP 40.
[6] WP 21.
[7] WP 35.
[8] WP 6 and 22.

Following software development best practices and peer review of publications can alleviate some of these concerns. For example, publishing software on a repository with a citable digital object identifier (DOI) creates a permanent version of the software, while others can modify the source code by adding a "branch" on a version-controlled repository. The primary developer can maintain control to any changes made to the primary source code so that if an error were introduced, the responsible party is identifiable through the version-control features.

Finding: Open source software enables community members to build on each other's labor and reduces duplication of effort. Especially for software that is broadly applicable, open source development can be more efficient and productive.

Finding: As more software is made open source, scientists will change how they work and how they evaluate each other's work, yet there is anxiety in the community about this impending change.

4.1.2 Collaboration and Inclusion

Some researchers report that open sourcing their software allowed them to broaden their collaborative network, advancing code capabilities in unexpected ways and improving their citation counts (Figure 4.1). All these factors enhance the software's impact on science (WP 7, 44). Others believe that OSS led to a broader collaboration network for the developer (WP 7).

Finding: Open source software can lead to a broader user and developer base for software, often fostering collaborations that are beyond what the original developers envisioned.

4.1.3 When to Open Software

Researchers sometimes find it difficult to determine exactly when to transition software to open source. Many science codes develop in stages, beginning as a research code that develops into a production code. For example, single-use software may transition to analysis software, or to simulation and model software. Scientists initially develop a research code to solve a specific problem or series of problems. On a time frame that can be much longer than a typical grant-funding period, scientists continue to develop and test the software, applying it to an increasing number of problems. During this period, the scientist may gain a set of friendly users (by-request access) who further test the software on a wider set of applications. This allows the developer to both fully test the software and also have time to publish a series of papers (return on investment) prior to releasing. Open sourcing the software occurs after this, sometimes lengthy, development/testing phase. Some express concern that transitioning to an open source requirement within a grant period may not allow developers the time to fully test the code or complete their own scientific findings with the code before releasing it to the broader community.[9]

In other cases, especially for software focusing on data mining and management (e.g., AstroPy), open sourcing the code from inception has produced not only a broad user group, but also a broad developer team. The broad developer base avoids duplication of effort and improves interoperability.

Conclusion: For many software projects, open sourcing the code from inception is ideal. For others, a period to verify and validate the code in a research mode may be a better approach.

4.1.4 Transparency and Reproducibility

Strong drivers for open data and OSS are the principles of reproducibility and replicability. Publications with only natural-language descriptions of methods, algorithms, and code implementation may be insufficiently

---

[9] M. Kunz, 2018, "My (Biased) Take On: Experiences and Challenges in Open Source Policies," presentation to the committee on January 17, 2018.

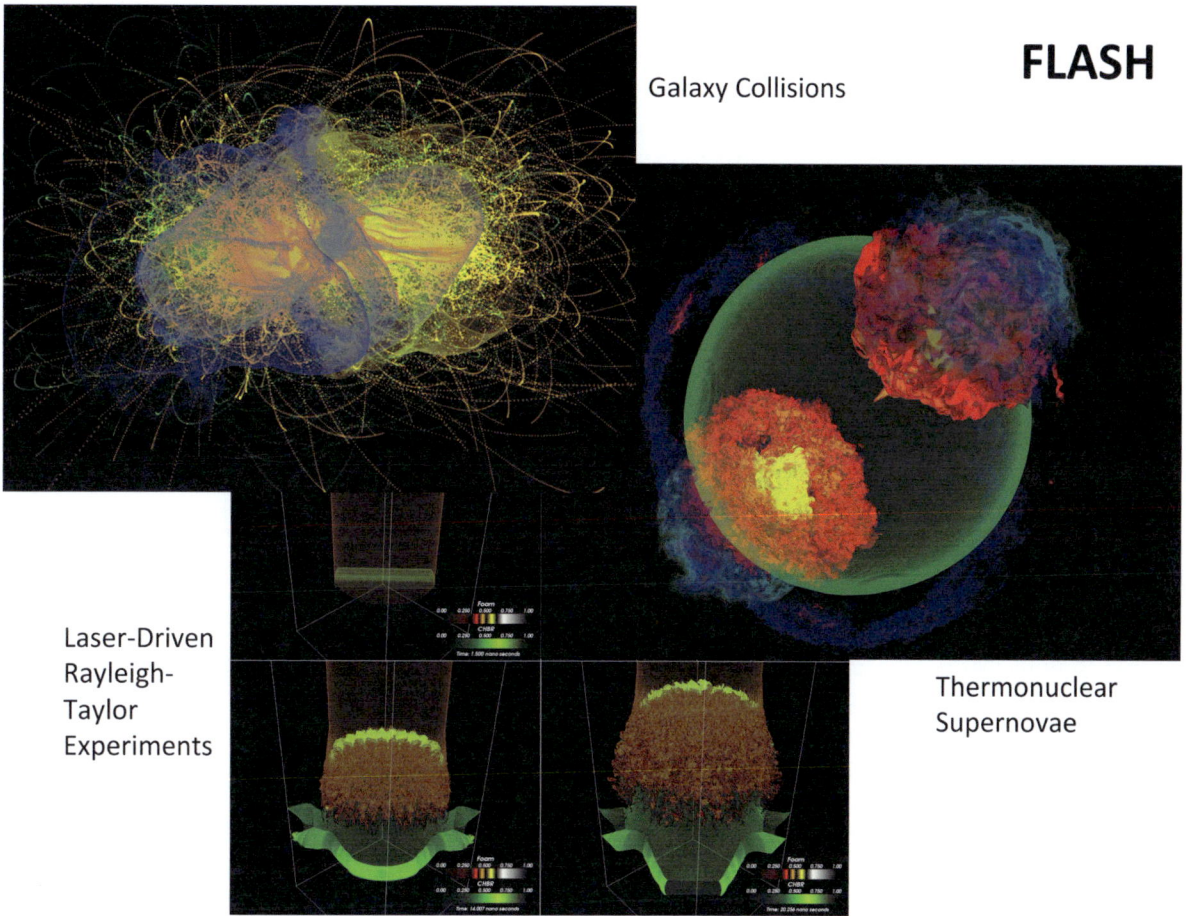

FIGURE 4.1 Examples of output from FLASH, a publicly available multiphysics, multiscale simulation code that has been used for applications in astrophysics and laboratory physics. SOURCE: Flash Center for Computational Science, http://flash.uchicago.edu.

reproducible and are sometimes misleading. Ince et al. (2012)[10] offer two examples. In the first example, data of global temperature anomaly released by the U.K. Meteorological Office included some errors that could have been discovered earlier if the software that processed the data had been open source. In the second example, a comparison of nine implementations of the same seismic data-processing algorithms revealed differences attributable to software errors. Ince et al. assert that ambiguity in descriptions using natural language are unavoidable, citing several works from software-engineering research literature. Even if description ambiguity could be eliminated, software errors are unavoidable. Some estimates report that 1 to 10 errors per thousand lines of code are typical.[11]

Strict reproducibility (the ability to reproduce results with the same data and code) requires OSS and open access to all the needed metadata including initial conditions, data inputs, libraries, compiling requirements,

---

[10] D.C. Ince, L. Hatton, and J. Graham-Cumming, 2012, The case for open computer programs, *Nature* 482(7286):485.

[11] B. Boehm, H.D. Rombach, and M.V. Zelkowitz (eds.), 2005, *Foundations of Empirical Software Engineering: The Legacy of Victor R. Basili*, Springer, New York.

computing environments, and so on. See, for example, Donoho et al. (2009)[12] and Nosek et al. (2015).[13] This can be difficult to achieve, especially in high-performance computing applications that require large portions of supercomputer facilities. It further requires careful bookkeeping to ensure that everything that is needed to produce the exact published result is archived. Reproducibility of complex computational environments can be facilitated with new technologies, such as containers. Tools such as Docker, Vagrant, and VirtualBox help researchers re-create a computational environment from a list of requirements. These new technologies are making reproducibility easier and are reducing barriers for building on past work. Yet, being able to reproduce a result with a specified metadata and a given code does not ensure that the results or code are correct. Replication requires that researchers use independent methods to achieve consistent results, verifying the scientific findings. Many communities already have a strong replication tradition, where trust in any scientific result is built when multiple codes achieve results that demonstrate consistent behaviors. By requiring multiple codes to achieve the same scientific finding, replication reduces the impact of individual code errors or numerical issues.

At least one journal, the *American Journal of Political Science* (AJPS), has adopted a rigorous reproducibility standard, including verification of each compendium by an independent organization. It is a costly process (approximately $1,000 per article), and fewer than 300 papers are published each year, with a 90 percent rejection rate, but AJPS's reputation has only increased since a adopting the policy, a sign of readers' trust in independently verified reproducible results.[14] While this type of compliance checking is not easily scalable to thousands of papers per year, journals are moving in this direction.

Reproducibility requires full transparency, enabling scientists to better independently replicate results. It is often difficult to fully describe in a publication the methods that were used with sufficient enough detail to reproduce the results. This is rarely intentional, but often details are missed in the capturing of the method in a publication. By opening software, scientists can more readily do code/method comparison, allowing for replication. OSS facilitates broader code review and builds trust among scientists and code projects.

> **Finding:** Reproducible research requires both open source software and good metadata, including initial conditions, data inputs, libraries, compiling requirements, computing environments, and so on.

> **Finding:** For many research projects, reproducing all results is neither tractable nor does it ensure that the results are correct. Replicability is also important. However, the transparency provided from open source software can foster and improve both replicability as well as reproducibility.

The committee found through presentations and discussions that the community has concerns regarding how OSS policies could potentially be exploited to "scoop" results. Some efforts have been made through licensing to mitigate concerns about first publication of results, but these licenses do not adhere to currently popular open-source definitions. Additionally, the publication of OSS would ensure that appropriate credit is given by providing a traceable record of the software publication. Making reproducible research a community norm will more likely be accomplished through incentives such as badging and journal or funder policies. NASA SMD may want to support that norm while ensuring fair credit and reasonable protection for researchers.

4.1.5 Institutional Challenges

OSS requires a change not only in the attitudes of scientists, but also in institutions. Some universities and companies tend to protect their intellectual property and may be reluctant to support OSS or may use restrictive licenses that could be in conflict with a future NASA policy. Some institutions, often for legitimate legal reasons, have developed a cumbersome system to release software as open source. Including an OSS condition at the

---

[12] D. Donoho et al., 2009, Reproducible research in computational harmonic analysis, *Computer Science and Engineering* 11(1):8-18, doi: 10.1109/MCSE.2009.15.

[13] B.A. Nosek et al., 2015, Promoting an open research culture, *Science* 348(6242):1422-1425.

[14] W.G. Jacoby, 2018,"The Replication and Verification Policy at the *American Journal of Political Science*," presentation to the committee on January 18, 2018.

grant proposal stage could encourage institutions to streamline, where possible, their approval processes, so as to meet funded contract or grant obligations. Institutions that receive grants from NASA are required to return annual reports on research progress, which may include progress on data and software development and release. The program manager overseeing the grant authorizes each year's funding increment. If a report was unable to show the release of software in a timely manner, this could affect the annual funding authorization or proposals for future grants, which may ask for documented sharing practices.

NASA's internal processes can also be cumbersome. One civil servant described how it took her "five months and 38 pages of paperwork to release 217 lines of non-sensitive code" (WP 1). The NASA Technology Transfer Program has worked hard to streamline this process,[15] and some NASA researchers have noted the improvements. At the same time, the chief information officer (CIO) for NASA Headquarters is encouraging a general approach.[16] The code.nasa.gov website outlines steps for scientists at NASA centers to follow in releasing software and already provides access to hundreds of software packages. It states, "Depending on the number of projects being assessed for release at any given time general workloads and backlogs, traversing the release process can take anywhere from 3 to 6 months."[17] Greater coordination between the CIO and Technology Transfer Office could clarify and accelerate the open source process. Since NASA employees also compete for funding, a NASA policy to release software as open source could encourage streamlined software release policies and procedures.

NASA's relations with contractors and grantees present other issues. The language used in NASA Federal Acquisition Regulations (FAR) conflates copyright ownership and the ability to distribute copyrighted materials, bundles software into data, and explicitly discourages open source (WP 12).

Conclusion: Achieving the goal of establishing a social norm of open source software requires altering the views of institutional management in addition to scientists.

4.2 EDUCATION AND TRAINING NEEDS

Education underpins all efforts to move the NASA science community toward acceptance of OSS. Software development experience varies widely across and within scientific disciplines, and the education component of any policy implementation needs to account for these differences. For example, in a 2015 informal survey of 1,142 astronomers, "only 8% of them report that they have received substantial training in software development. Another 49% of the participants have received 'little' training. The remaining 43% have received no training."[18] For example, some programs (e.g., ACCESS[19]) have projects led by professional software developers, but most SMD projects are led by scientists, who may not be as conversant with the latest coding standards or techniques. A move toward OSS is seen by many researchers as imposing extra requirements unrelated to or detrimental to the scientific quality of a project (WP 6, 9, 10, 15, 21, 22, 26). Clear communication with the research community about new requirements is essential for success. Training scientists in best software practices is critical to gaining the full benefit of OSS (WP37). Convincing them of the long-term benefits to their science is the best (perhaps only) way to gain their acceptance of an OSS policy. There are already strong community efforts to educate scientists about software development. Rather than replicating them, SMD could sponsor events such as face-to-face Software Carpentry[20] workshops, online classes,[21] or mentorship workshops for late-career scientists. For example, Software Carpentry workshops do not tell participants how to write their code, but they teach them how to use version control software and how to create a citable persistent identifier (e.g., DOI) for their software using Zenodo.[22] Figure 4.2 outlines

---

[15] D. Lockney, 2018, "NASA Software Release," presentation to the committee on January 18, 2018.
[16] Jason Duley, discussion with the committee, February 16, 2018.
[17] See https://code.nasa.gov/#/SRA.
[18] I. Momcheva and E. Tollerud, 2015, "Software Use in Astronomy: An Informal Survey," arXiv:1507.03989v1.
[19] Advancing Collaborative Connections for Earth System Science (ACCESS), https://earthdata.nasa.gov/community/community-data-system-programs/access-projects.
[20] See http://software-carpentry.org, a nonprofit that focuses on teaching people to code.
[21] For example, https://codecademy.com/ and https://datacamp.com/, for profit companies that have online classes teaching coding.
[22] Zenodo is an online repository for research output. Zenodo was created by Open Access Infrastructure for Research in Europe (OpenAIRE) and CERN and is supported by the European Commission, https://zenodo.org/.

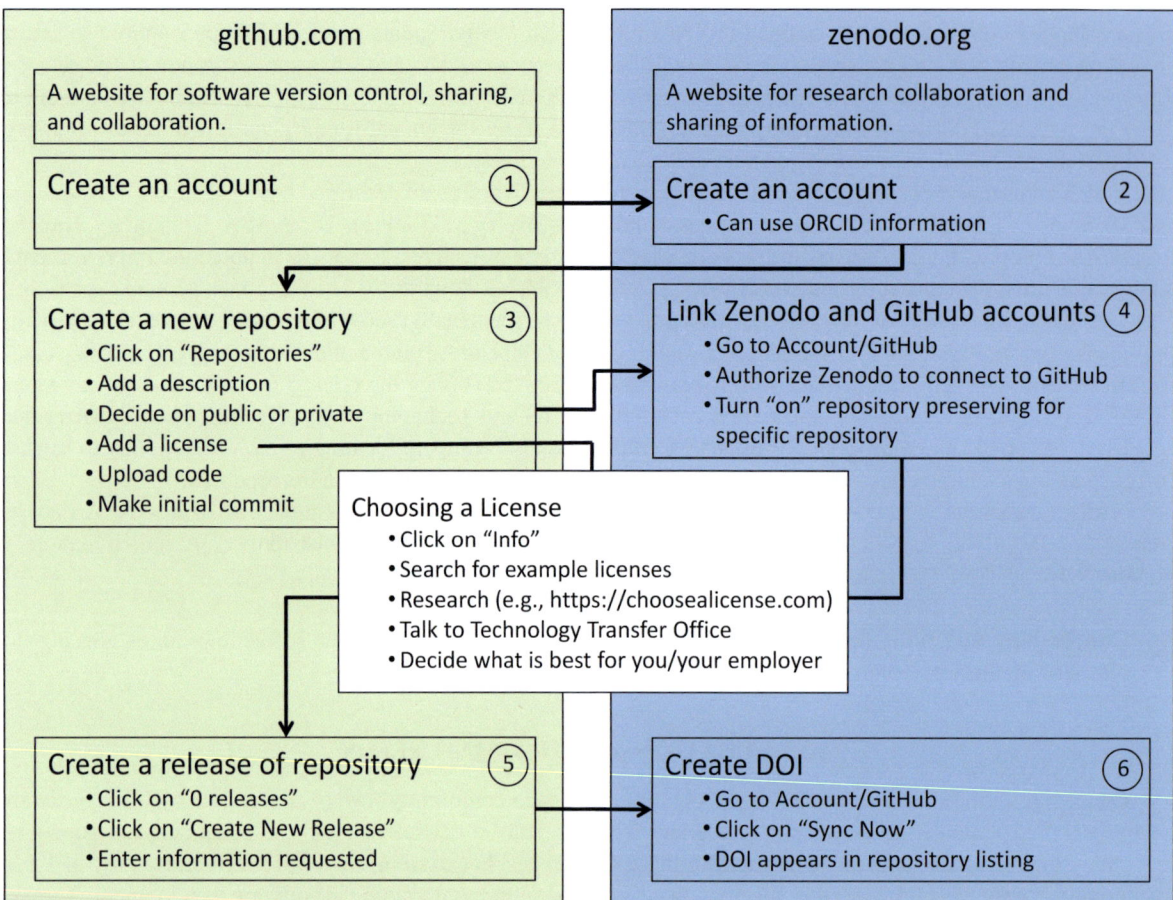

FIGURE 4.2 Process of releasing open source software on GitHub and assigning a digital object identifier (DOI). NOTE: GitHub is a hosting service for software that supports version control and distributed collaboration workflows (it offers free as well as paid accounts). Zenodo is a general-purpose data repository originated at the European Organization for Nuclear Research (CERN). SOURCE: A. Ridley, 2018, creating-a-doi.pdf, https://doi.org/10.6084/m9.figshare.7015289.v1.

the basic process of how a software developer could take software and make it open source using both GitHub and Zenodo. While there are many steps, they are all relatively straightforward, with the most complicated one being choosing a license. Once this is done, publishing the code and getting a DOI that can be used in publications is simple. Additional topics for education include but are not limited to the following: (1) how to organize and publish software in a repository (e.g., GitHub); (2) how to choose and include a license; (3) what to consider before sharing software (university rules, export controls, etc.); and (4) other OSS best practices.

Finding: Community understanding, experience, and familiarity with open source software is essential to the acceptance of open source software as a tool to enhance scientific research.

Two community deficiencies were of particular concern: the lack of training in software development and scientific computing and the missing guidance on legal issues for scientists.

4.2.1 Modern Computing

Some committee members noted, from their own experience, that young scientists may not have adequate training in modern computing and software development and are likely to reuse existing code rather than create new code. This presents both advantages and disadvantages. Reuse can help a beginning programmer learn while creating software faster and more efficiently, but using these programming tools could circumvent understanding what a piece of code actually does. Science is increasingly dependent on software for analysis, but many scientists lack the training in software development best practices and this is a barrier to collaborations. This is a national trend that is likely due to a multitude of factors. Regardless of the cause, communication of this skill mismatch, and the fact that coding is becoming as essential as calculus to scientists, could motivate secondary schools and colleges to include software development best practices in their curricula for all science, technology, engineering, and mathematics (STEM)-bound students. OSS provides a way to educate and train new talent (WP 40).

4.2.2 Guidance on Legal Issues

Determining whether something is export controlled is nontrivial, but if an OSS policy is to succeed, it will have to become trivial (WP 5). Much of the Earth and space science community lacks basic knowledge about intellectual property law, contractual wording, software licensing, institutional policies, and export control, because this knowledge has not been required for their research. While most research is not export controlled, checking for sensitive technologies is part of due diligence when releasing software. In some cases, safely releasing software is simply a matter of finding the correct legal resources within a scientist's institution. For smaller institutions or independent scientists, guidance may be hard to find. Implementing a policy without ensuring the necessary access to legal resources for NASA-funded scientists would impose a burden (WP 8) and create inequalities in access to legal advice (WP 15).

NASA has several sites for releasing software that begin to address some of these concerns, but they are not well known. The code.nasa.gov website gives a step-by-step procedure for NASA employees to release OSS (Section 3.1.1). At software.nasa.gov, any user can request software, but the release type enforces export controls. WP 5 documented the difficulty one NASA employee experienced while trying to understand whether the software he wanted to release was export controlled. Better informing the community of such sites, and further developing them, could provide a centralized distribution and education facility for SMD-funded OSS.

Finding: The scientific community's lack of knowledge regarding the legal issues is a currently a barrier to releasing open source software.

Conclusion: Training and education on legal issues and open source software best practices will improve acceptance of any new requirements.

Recommendation: NASA Science Mission Directorate should initiate and sponsor programs to educate and train researchers in open source best practices. Topics could include, but are not limited to, export controls, licensing and intellectual property, workflows, and software development. These resources could be made available to the community via in-person trainings as well as webpages, screencasts, and webinars.

4.3 FUNDING AND EFFORT NEEDS

It is important that NASA consider the cost of a future OSS policy against the impact on innovation and scientific productivity and act to minimize new non-science-related burdens for science proposals, including training requirements when implementing any OSS policies.

4.3.1 Funding

Despite many community members expressing support of OSS because it has advanced their research and careers, as discussed above, others are concerned that releasing OSS may disadvantage those scientists who have spent time on the code development (WP 15, 21, 26, 29, 32). Because funding and career advancement are tied to traditional scientific output (e.g., publications) rather than code development, some researchers feel the need to maintain closed software in order to have an advantage when securing funding. Similar concerns were expressed when NASA adopted an open data policy.

There is also a concern that NASA's current funding mechanisms do not support the time it takes to document, test, and maintain robust, reproducible, and reusable software. Although open source does not require these elements, they are necessary to maximize its benefits, and sometimes are implicitly expected (WP 11, 15, 37). Many believe that making the analysis code available as is (i.e., without documentation) is unlikely to enhance scientific productivity (WP 16 and 21).

Another concern of the community is that any mandated OSS policy without associated institutional and financial support would be untenable to scientific code developers (WP 6, 7, 12, 15, 18, 19, 24, 31, 41, 44). If there are increased costs to software development imposed by NASA without clear funding channels, developers may choose to work on non-NASA projects with funding from other sources. Additional funds to open source codes will help facilitate more OSS. The cost may be prohibitive for legacy codes (WP 32). Maintaining open source libraries and software requires the allocation of resources (WP 7, 8, 10, 11, 22, 33, 39, 44).

Finding: Making code open source is valuable, but NASA Science Mission Directorate will need to consider the additional costs to the developers and to NASA. For many codes, additional funding will be needed to document, test, and maintain robust, reproducible, and reusable software. While the benefits of open source software start at the simplest step of openly sharing software, to achieve the full benefits of open source software, adequate funding will be required.

4.3.2 Effort

OSS development enables software reuse, avoiding duplication of effort (WP 3, 26). Open code also encourages a larger developer base, which can contribute to maintenance, verification, and validation of the code (WP 3, 16, 34, 39).

OSS does require time and effort by both researchers and institutions to identify the appropriate license, implement adequate development practices, and so on. Finding and implementing all the open source components used by software can be difficult and time consuming, complicating future projects (WP 10, 13, 43). An example complication is when software projects are funded by multiple agencies with different open software policies. Another example is that the use of commercial-off-the-shelf (COTS) software in an era of open source may cause confusion unless the policy is very clear. There is concern that new policies may preclude the use of COTS tools to support research, which could have a negative impact (WP 13).

In addition, many scientists feel compelled to "clean up" and document code before open sourcing (WP 6, 10, 18, 40). Once software is open, additional work is required to provide support and enhance an existing code. Although this could lead to higher-quality codes, there is some concern that scientific productivity and innovation could be negatively impacted by this extra work.

Making single-use code open source may not be worth the extra effort (WP 6, 13, 18, 21). Single-use codes can evolve into production codes with broader applications, at which point benefits from open source development could be larger and, for some software, delaying the time to open source may be warranted (WP 29).[23,24]

Finding: The perception that considerable effort is required for open source software may inhibit its adoption.

---

[23] P. Woodward, 2018, "Thoughts on Open Code Policy," presentation to the committee on January 17, 2018.

[24] M. Kunz, 2018, "My (Biased) Take On: Experiences and Challenges in Open Source Policies," presentation to the committee on January 17, 2018.

Conclusion: An incremental approach to open source software will allow researchers to adjust to new requirements and minimize the impact on their scientific productivity.

Conclusion: Flexibility is needed in an open source policy.

Recommendation: Any open source software policy that NASA Science Mission Directorate develops should not impose an undue burden on researchers; therefore, any policy should be as simple as possible, and any mandates should be fully funded.

4.3.3 Support for Good Practice, Governance, Maintenance, and Infrastructure

The implementation of a NASA OSS policy necessitates a governance infrastructure to ensure software quality and security. This type of infrastructure requires a sustained financial investment with clear roles and responsibilities for staffing and participation (WP 37). Lessons learned from the implementation of NASA's open data policy illustrate the importance of design and planning of a supporting infrastructure for open software prior to any policy implementation.

As research scientists publish their software and use community resources, such as tools and libraries, the software needs to be managed, discoverable, and, where appropriate, maintained in a centralized (as much as possible) repository (WP 30). There are existing NASA data resources that publish sensitive software[25] and general open software[26] that could be leveraged for new OSS initiatives. In addition, NASA can encourage the open source development model described in Section 2.5.

Recommendation: NASA Science Mission Directorate should support the infrastructure, governance, and maintenance of a healthy open source community, taking advantage of existing community resources to the greatest extent possible.

4.3.4 Support for Community Software

NASA science is enabled and dependent on open source, community-developed software, from fundamental packages such as NumPy, which performs standard array operations, to domain-specific ones such as AstroPy, which enables many common astronomical calculations. The continued existence, active development, and widespread adoption of open source libraries by many federally funded projects demonstrate the power of community-developed software. Some successful open source projects are sustained by organizations (often private companies) that gain training for future employees and libraries for existing employees to use, and community goodwill. However, most open source libraries are currently maintained by volunteers and lack clear sustainability plans. This lack of institutional support, dedicated full-time developers, and dedicated funding for these libraries represents a major vulnerability of the basic infrastructure of NASA science. Muna et al. (2016)[27] described several options for securing this infrastructure. NASA could allocate professional software developers to these projects, could encourage NASA employees to participate in the projects, or could provide funding directly to the projects to hire the needed developers. An example of a government agency supporting open source libraries and working with the existing open source community is the Department of Defense (DOD) Defense Advanced Research Projects Agency (DARPA) Open Catalog.[28]

Finding: NASA science already depends on open source, community-developed libraries.

---

[25] See https://software.nasa.gov.
[26] See https://code.nasa.gov.
[27] See https://arxiv.org/abs/1610.03159.
[28] See https://opencatalog.darpa.mil.

Finding: Science built on open source software is most effective when it has a strong, coordinated, and active community.

Conclusion: Community software provides substantial value to SMD-funded researchers, and NASA's recognition and support of these software projects is vital.

Recommendation: NASA Science Mission Directorate should support open source community-developed libraries that advance NASA science.

4.4 ENABLING CREDIT AND CAREER ADVANCEMENT

Community members are concerned that making software open source makes their intellectual property vulnerable to reuse without attribution (WP 6, 15, 20, 21, 42).[29,30] In the past, developers of OSS have struggled to obtain permanent positions in science (WP 44). Although new policies are being developed to ensure attribution (WP 7, 44) for these efforts, more work is possible to protect the careers of open source developers.

Moving to OSS can help with career advancement, especially outside academia. Examples exist where software is held by an institution, and, when a scientist leaves, this software stays at the institution, often with no viable support mechanism. OSS would allow the scientist to continue building a career using the software. Open software protects scientists from companies, universities, or laboratories interfering with their work when they move between institutions. But despite recent improvements, the perception remains of a lack of respect and appreciation in academia for software development. NASA investments can help change this perception. Recognizing science software as a critical element for innovation and investing through a long-term software development and maintenance program may elevate the status of the software developer. Academic credit for OSS is important as well, especially through formal citation in the scientific literature and encouraging the acceptance of OSS as a community norm. As described in Chapter 3, some journals are beginning to require that the software used to create a research result be made available. Ideally, this would facilitate formal citation using a persistent identifier such as a DOI. Software citation can promote both scientific reuse and formal recognition of software developers, but software citation practices are not yet firmly established. Several journals, including *SoftwareX*, the *Journal of Open Research Software*, and the *Journal of Open Source Software,* provide a means to publish peer-reviewed "software papers" that include release of the described software. This is appropriate for some but not all software; yet, much like data, any software used to produce a published scientific result needs to be cited. Data citation principles and practices are becoming more firmly established,[31,32] and similar principles for software citation are developing, but practices are not as mature.[33] Furthermore, citation is only one method of addressing credit. Projects such as depsy,[34] Project CRediT,[35] and others have explored new ways to understand different types of contributions and the general concept of "transitive credit." Note that changes or adaptations to software by someone other than the original author are more easily done on software openly released on GitHub or similar repositories, where the original software is clearly and persistently identified, and each individual's contributions or separate work are identified through "commits."

NASA can encourage the further development of appropriate credit schemes by formally recognizing software contributions in grant and contract proposals, interim and final project reports, and hiring and performance

---

[29] M. Kunz, 2018, "My (Biased) Take On: Experiences and Challenges in Open Source Policies," presentation to the committee on January 17, 2018.

[30] W.R. Hix, 2018, "One Programmer's Experience and Challenges on Open Source Policy Decisions," presentation to the committee on January 17, 2018.

[31] Data Citation Synthesis Group, 2014, *Joint Declaration of Data Citation Principles,* M. Martone (ed.), Force11, San Diego, https://doi.org/10.25490/a97f-egyk.

[32] J. Starr, E. Castro, M. Crosas, M. Dumontier, R.R. Downs, R. Duerr, L.L. Haak, M. Haendel, I. Herman, and S. Hodson, 2015, Achieving human and machine accessibility of cited data in scholarly publications, *Peer Journal Computer Science* 1:e1-, http://dx.doi.org/10.7717/peerj-cs.1.

[33] A.M. Smith, D.S. Katz, K.E. Niemeyer, and FORCE11 Software Citation Working Group, 2016, Software citation principles, *Peer Journal Computer Science* 2:e86, https://doi.org/10.7717/peerj-cs.86.

[34] *Nature* article describing depsy.org at https://www.nature.com/news/the-unsung-heroes-of-scientific-software-1.19100.

[35] See http://casrai.org/credit.

review processes. The goal is to foster the norm of OSS by explicitly documenting and registering contributions and recognizing the value of those who contribute.

Conclusion: In order to recruit, retain, and support the skills base, software development efforts need to be rewarded with academic credit (e.g., grants, publications, citations, fellowships, and prizes).

Recommendation: NASA Science Mission Directorate should foster career credit for scientific software development by encouraging publications, citations, and other recognition of software created as part of NASA-funded research.

5

Policy Options and Recommendations

The committee outlines a selection of policy options below, including both incentives and mandates for NASA's Science Mission Directorate (SMD) to consider. Based on the charge to the committee and discussions with NASA officials, the committee operated under the assumption that SMD will transition to a greater level of openness in accordance with federal policy (see Chapter 1). It is important, therefore, that NASA ensures that the transition helps advance science, foster collaboration, and generally advance NASA goals (see Section 1.2). The committee believes that the best way to achieve this is to work toward a cultural norm of robust, open source software (OSS) development and maintenance. This will not happen overnight and will require ongoing strategic investment.

SMD does not currently have division-wide policy regarding software publication, distribution, or licensing. As described in previous chapters, each software type may have its own legal requirements, and raise different policy issues and concerns from the community on the value and practicality of openness. Correspondingly, existing NASA software varies across a spectrum of openness. Each division may have its own policies, and the chief information officer (CIO) and Technology Transfer Office may differ on matters of OSS release. SMD may choose an overall level of openness for new software produced by and for the directorate in the long term, but individual programs and software types will advance toward greater openness at their own rate and through different means. The options below can be considered a sort of toolbox to draw from and help move the community toward greater openness while recognizing that different disciplines and code types will have different requirements and transition at different rates. Incentives will help to move the community norms toward greater openness regardless of whether mandates are eventually implemented. Overall, the committee believes that there will need to be a combination of different incentives in place and transition to mandates only as appropriate.

Recommendation: NASA Science Mission Directorate should consider a variety of policy options depending on discipline and software type and transition to greater openness over time.

With all that in mind, the committee presents the following three open source policy options at a basic level:

Option A: Continue status quo.
Option B: Incentivize openness.
Option C: Mandate openness.

Continuing the status quo (Option A), allows individual SMD programs to determine whether they and their research communities are interested in moving toward open source. Ideally, regardless of what NASA decides, as accepted research metrics begin to routinely include a measure related to software impact, OSS will gradually become more common at least in some disciplines. Option A could eventually lead to OSS being required or a de facto norm in some areas, but in others it would remain unusual. This is the least strategic option and is unlikely to bring much change or fully realize the value of OSS.

SMD could accelerate the move toward openness, especially in programs less inclined to it, by offering specific incentives to investigators (Option B). This could be the long-term policy of SMD or could be a step on the way to some level of SMD-mandated openness (Option C). These incentives can also contribute to fostering an OSS culture and norm of sharing. Section 5.4 describes how incentives and mandates can be combined for different types of software.

5.1 POLICY OPTION A: CONTINUE STATUS QUO

As stated earlier, SMD does not currently have division-wide policy regarding software publishing, distribution, or licensing. Option A continues to allow individual NASA programs to determine whether they and their research communities are interested in moving toward open source. Some programs and modeling centers have taken steps toward openness, but there are no SMD coordinated open-source requirements, education efforts, or trainings. Several examples of existing policies for parts of SMD are discussed in Chapter 3. The community concerns raised in Chapter 4 would mostly remain unaddressed. This option could lead (and may have already led) to inequalities in access to results if some, but not all, programs mandated OSS. Without guidance from NASA, others (including publishers who are increasingly requiring OSS) will determine the course of OSS policies.

Option A also leads to approaches customized to certain communities. For example, some large modeling centers, such as NASA Community Coordinated Modeling Center (CCMC), have what might be called an "open-to-run" approach. They host software packages and run them on in-house hardware, possibly on thousands of CPUs for months at a time. This allows outsiders to run software at the center with cases and configurations of their choice, but under various restrictions. Usually, the software cannot be taken elsewhere. Sometimes the source code is viewable; sometimes not. This ad hoc approach may be a pragmatic compromise, because the software cannot be practically run on other systems. It may also be a way to provide some protection for the investigators developing the software while satisfying some level of research transparency. Investigators may have some ability to reproduce results and learn from the software depending on the restrictions, but this is not OSS.

5.2 POLICY OPTION B: INCENTIVIZE OPENNESS TO ACCELERATE THE CHANGE

Option B uses incentives that preserve community interests while moving to OSS. The goal of this option is to build trust while working toward making openness a community norm. With mandates absent or delayed, community pressure toward openness would naturally increase as investigators compete for the incentives.

As described in Chapters 3 and 4, some policies, such as the compulsory inclusion of a data management plan, was initially not well understood and was seen as a burden to most investigators, even those who do not produce any data. As education efforts and online resources increased, data management plans became a normal section to all proposals. The archive of widely used data at NASA archive centers, rather than individual investigator's websites, has improved the quality, availability, and usability of NASA investigator-produced data. NASA's Living with a Star (LWS) program offered researchers funds to make widely used data available to the general community and was better received than unfunded mandates. The implementation of an OSS policy could learn from these experiences, through implementation of options given below.

Success with Option B depends on the allocation of adequate resources. Incentives within the current budget that lead to reduction of research funds will be less accepted by the community. There may be a delay in the scientific return from research funding. Over the long term, however, motivation to move toward more openness is likely to provide a net benefit to science, as more researchers take advantage of the opened software. Careful consideration and guidance to proposal review panels would be required to award incentives to those proposals with software that is more likely to be reused. This option also recognizes the full cost of community software development.

Because openness incentives may be applied at different rates and perhaps be absent at other governmental agencies, some researchers may not participate, possibly gaining a research or career advantage over those who devote time and resources to opening their software. For example, if one program manager or agency called for OSS and another did not, a researcher might target proposals to programs that allowed them to delay releasing their software.

The five incentives identified by the committee are the following:

B1. Funding for new proposals specifically addressing an OSS need
B2. Funding augmentations or components of proposals
B3. Piloting the use of software management plans (SMPs) in some programs
B4. Supporting open source libraries and infrastructure software
B5. Offering a prize for exemplary contributions to OSS in the NASA science community.

One or more of these elements could be adopted as part of Option B. Table 5.1 summarizes how these may apply to the different software types presented in Chapter 2.

5.2.1 Option B1—Funding for Full Open Source Software Proposals

With Option B1, SMD or its divisions allocate funding, either in existing or new grant programs, to open existing software with community reuse potential or replace it with functionally equivalent OSS, to develop new OSS, to maintain existing OSS, or to extend community open source libraries and frameworks. In this case, the proposal provides a software management plan (see Box 5.1) that describes what the software does, its value and user community, the needed software and documentation updates, test suites, licensing, any other legal issues, and a plan for long-term sustainability. The funding could be delivered as contracts or cooperative agreements to allow full oversight with milestones and deliverables.

Similar to the implementation of NASA's data management plans, a corresponding NASA website could explain what is required for SMPs and provide suggestions and examples to ensure less-experienced researchers are not disadvantaged. Online tools (e.g., https://dmptool.org) could be developed to further help investigators.

Pros: Option B1 provides direct funding for scientists to develop OSS solutions. It allows for a prioritized approach, as evaluation panels of community experts decide which software to support. It recognizes the cost of community software development and it creates or builds scientific software projects that other researchers can reuse, accelerating science.

Cons: Option B1 could delay scientific returns within programs implementing it, as scientists spend time to gain experience and familiarity releasing software. It opens only some software, and it may provide a disincentive for groups who do not win funding to open their software. It may not fully recognize the long-term costs of maintaining the software.

5.2.2 Option B2—Optional Proposal Open Source Add-On

With Option B2, SMD provides an opportunity to add additional pages to scientific research proposals in existing grants programs to justify additional effort and funding to open software from the project and to provide an SMP. The review panel and program manager evaluate add-on proposals and allocate funds to the best of them. Unlike Option B1, the OSS management plan and funding is an augmentation to the scientific proposal.

The pros and cons are similar to those for Option B1, with the difference that there may be situations where the scientific merit of the proposal is not rated high enough for support, but the open source add-on is seen as of significant value.

TABLE 5.1 Policy Recommendation Summary by Software Type

	A: Continue Existing Policy	B1: Incentivize with New Opportunity	B2: Incentivize with Proposal Add-on	B3: Pilot SMP for New Development	B4: Support Existing Open Infrastructure	B5: Prize	C: Mandate Open Source	Remarks
Libraries	✗	✓[a]	✓	✓	✓	✓	Immediate for new software or additions to existing OSS	
Single-use software	✓	✗	✓	✱[b]	✗	✓	Last priority; most difficult to implement	A greater priority for journals implementing reproducible research standards.
Analysis software	✗	✱	✓	✓	✱	✓	New tools ASAP ~5 years	
M&S software	✗	✱	✓	✓	✱	✓	Variable	
Frameworks	✗	✓	✓	✓	✗	✓	Immediate for new; as possible for old	
Sensor software	✗	✓[c]	✗	✓	✗	✓	Immediate	Include in new AOs as soon as possible.
Infrastructure software	✱	✗	✱	✱	✱	✓	When necessary for science and to ensure competition and good practice	

[a] Yes, for maintaining current libraries and creating major new libraries.
[b] Items with an asterisk (✱) indicate that the policy may be applied in some programs before others.
[c] Yes, for high priority, general-use legacy pipeline.

NOTE: General approach of moving toward a full SMD-wide OSS policy based on software types described in Chapter 2 (see Table 2.1). The table indicates which policy options may be appropriate for each software type and the recommended time frame for moving to mandated openness (Option C). ✓ means that an option can be implemented right now; ✱ means that it might be implemented in some programs before others, and likely with lower priority; and ✗ means that the option is not appropriate for that code type. An ✗ might appear because going straight to a mandate is recommended for that software type (e.g., sensor software), or because a more comprehensive solution involving OSS and other requirements would be necessary (e.g., single use). More detailed descriptions and recommendations are given below.

> **BOX 5.1**
> **Software Management Plans**
>
> Software management plans (SMPs) with specific requirements are already required as part of new NASA mission and supporting infrastructure proposals (e.g., SWE-102[1]). Within this report, an SMP is a document that concisely describes all new software produced during a research project; intellectual property or legal issues that might arise; and how the software will be developed, shared (or not shared), and archived during the project.
>
> Most of the policy options described will require SMPs, which would include the following:
>
> 1. A description of software that will be produced during the funded research and other components used (OSS, COTS, SaaS, etc.). It may be appropriate to separate software into categories for this discussion and following information.
> 2. Version control processes.
> 3. The license to be used (if released).
> 4. Any legal issues involved in the software components (e.g., export control and ITAR restrictions).
> 5. Long-term sustainability plan (if appropriate).
>
> SMPs that involve releasing software would also address the following:
>
> 1. Analysis of risks in releasing software (e.g., security, intellectual property, and licensing).
> 2. A software release schedule.
> 3. Testing and integration processes.
> 4. Software distribution and archiving.
> 5. Documentation provided to the community.
> 6. Methods for incorporating contributions from the community into the software and communicating those contributions.
> 7. Evidence of understanding and commitment to established open source development practices.
> 8. Budget description for software development, documentation, distribution, support, publication, and maintenance.
>
> ---
> [1] See https://swehb.nasa.gov/display/7150/SWE-102+-+SW+Development-Management+Plan.

5.2.3 Option B3—Pilot Software Management Plans

With Option B3, specific programs within SMD begin to require SMPs for scientific proposals containing substantial new software development. Requiring an SMP would not mandate openness, but it could gradually expand existing policy and impose more specific requirements over time. The goal is to gradually develop an effective policy by identifying different approaches to making software more open and responding to community feedback.

Where researchers receive NASA funds to deliver OSS, they also need to provide evidence of OSS practices. The SMP could be added as an evaluation criterion during proposal reviews.

Pros: An SMP would provide an opportunity for the research community to document and be evaluated on their software practices and for NASA to learn from the experience. It will allow reviewers and program managers to understand how knowledgeable and experienced the proposers are at OSS development.

Cons: Option B3 imposes an additional requirement that researchers must adhere to and evaluators must consider. If implemented rapidly in scientific communities unfamiliar with OSS, innovation and science could suffer due to either inexperience making software open or the additional time spent by researchers on software.

It is unclear what implications successful proposers would face if their stated SMP goals were unmet, except by evaluating previous practices, which would only apply to previous OSS funding from NASA.

5.2.4 Option B4—Support Open Source Libraries and Infrastructure Software

With Option B4, SMD uses existing funding mechanisms or allocates SMD employee time to support and adopt open source libraries and infrastructure software that are widely used in NASA-funded research. NASA support of software will demonstrate its commitment and promote the careers of scientists who spend time developing and improving these libraries. Because it is focused on broadly useful software, share-in-savings contracts might be one useful approach in this option.[1]

Pros: Option B4 could improve community software quality and generate savings for NASA as a whole. With agency support, community software will increase in visibility and value, which will help the careers of researchers who develop them.

Cons: Some software of this type currently exists without dedicated NASA funding.

5.2.5 Option B5—Create an Annual Prize for the "Advancement of OSS Development and Impact"

With Option B5, greater recognition for scientists who create quality OSS would enhance their career advancement. A NASA SMD award or prize could provide some recognition and visibility for the importance of OSS. This might be similar to NASA's current "Software of the Year" award,[2] but the focus would be on open source. The prize would recognize how OSS provides value to NASA. The prize could be judged on a number of criteria including, but not limited to, code contributions, software reuse or extension, training, and advocacy, as documented by application materials or testimonials and letters of nomination. NASA could consider partnering with scientific professional societies to present and administer the prize.

Pros: The prize provides an incentive and recognition for scientists who create quality OSS, while highlighting the importance of NASA SMD OSS resources.

Cons: The prize will require resources and take time away from other activities, as SMD needs to create the prize, publicize it, organize a review committee, review applications, and make a selection. Awards and prizes are not nearly as effective at advancing scientific careers as funded proposals unless they are extremely prestigious.

5.3 POLICY OPTION C: MANDATE OPENNESS

Under Option C, SMD decides that, by a certain date, software created through SMD funding will be open source, with only few, strongly justified exceptions. Mitigating the concerns raised in the prior chapters requires a period of transitional activities that may vary in timing and implementation by program and software type (Table 5.1) and sustained strategic investment.

Pros: An OSS mandate would be the surest and quickest way to increase the transparency of NASA science and to satisfy the recent Office of Management and Budget (OMB) memorandum "Federal Source Code Policy: Achieving Efficiency, Transparency, and Innovation through Reusable and Open Source Software"[3] and NASA CIO response to "encourage vendors to use open source technology wherever possible."[4] It could potentially enhance NASA's national and international reputation as a leader in open science. Experience with open data policies suggest than an OSS mandate could drive other agencies, both nationally and internationally, to enact similar policies, resulting in more OSS, thereby benefiting NASA researchers.

---

[1] Share-in-savings contracts are where "the Government awards a contract to improve mission-related or administrative processes or to accelerate the achievement of its mission and share with the contractor in savings achieved through contract performance," https://www.law.cornell.edu/uscode/text/10/2332.

[2] NASA Software of the Year Award, https://partnerships.gsfc.nasa.gov/internal-inventors/awards/nasa-technology-awards-and-incentives/#softwareoy.

[3] See https://sourcecode.cio.gov.

[4] See https://code.nasa.gov/NASA-M-16-21-OCIO-Memo.pdf.

Allowing flexibility in the timing and implementation of a mandate for different software types or programs could improve buy-in and understanding from the NASA research community, especially if NASA uses the time to implement training programs and some of the incentives under Option B. This flexibility in implementation would also allow costs to be spread over time and offers more flexibility in budgeting.

Cons: A rapid mandate would likely cause backlash from NASA-funded investigators. It would likely put a financial strain on NASA-funded investigators, especially those on soft money and could lead to a drain of people from the field. The response from the community may be to follow the letter but not the spirit of the policy. A mandate may also be the costliest option, requiring major enabling and sustaining infrastructure to enforce the mandate. The cost of implementing mandates could exceed the benefit for some software types. A mandate might also hinder collaboration with other agencies in creating OSS, notably the Department of Defense, which would have to provide permission and may have additional security concerns to protect controlled unclassified information and export-controlled information (WP 31[5]). Last, behavioral research suggests that positive incentives can effectively drive both individual and group behavior, so mandates are more effective once incentives have been established.[6]

Conclusion: Immediately mandating open source across all software types and in all of SMD could damage the NASA science enterprise.

If mandates are implemented, a transition period toward openness is necessary with specific activities to help level the playing field, provide training and resources, and ensure research continuity. In many cases, incentives will need to be in place before firm mandates can be implemented. The transition schedule may need to be modified based on the response to incentives as they are developed. Program managers will play an important role by tailoring their policy execution to their understanding of communities and culture across SMD, to incremental needs depending on software types, and to community experience with open science practices.

Conclusion: An incentive-driven transition period is needed before a comprehensive SMD open source software policy is implemented. Incentives and timelines will vary by software type and community experience.

5.4 TRANSITIONING TOWARD OPENNESS

As mentioned, SMD should consider a variety of policy options depending on discipline and software type and transition to greater openness over time. In Box 5.2, the committee outlines one example of a path to openness for a program or software type that moves to an OSS requirement in 3 years. The committee considers 3 years the minimum transition time applicable to only some software types or communities (e.g., new instrument code or pilot efforts such as NASA's Earth Science Directorate's ACCESS). Many programs or software types will transition more slowly because of different grant cycles, infrastructure availability, and general community readiness; but they will follow the same general path. Achieving SMD-wide openness means implementing this path for every program and for all software types. There will, however, be limitations. Some software cannot legally be open source; it may simply be too expensive to convert some legacy software; and different software will have different levels of maintenance (sometimes none). SMD will need to continually balance trade-offs and priorities while continually assessing how policies are meeting their goals. Early assessments are critical in policy implementation and establish a baseline for long-term assessments (e.g., reuse, publications, citations). This transition to a desired level of openness requires time and resources for training, software support and maintenance, and contributions to the overall software infrastructure. Introducing OSS requirements without strategic investment in software development and maintenance may not advance innovation and discovery and other goals.

---

[5] WP is used to reference the white papers submitted to the committee. See Appendix C for a full listing.

[6] E.E. Smith, S. Nolen-Hoeksema, B.L. Fredrickson, G.L. Loftus, D.J. Bem, and S. Maren, 2003, *Atkinson and Hilgard's Introduction to Psychology*, Wadsworth/Thomson Learning, Belmont, CA.

> **BOX 5.2**
> **Path to Openness Example**
>
> Three or more years prior to open source software (OSS) requirement:
> 1. Require a software management plan (SMP) for all proposals.
> 2. Offer competitive funding for opening existing software.
> 3. Review the program for existing software that needs to be open, and proactively approach investigators about opening their software. Consider and implement some of the incentives in Option B to begin to open software.
> 4. Begin to offer training and education programs on OSS creation, publication, use, and attribution at major conferences, webinars, and through other outreach programs. Provide a complete and detailed website for SMPs, NASA's OSS policy, a FAQ, and OSS guidance for potential proposers.
> 5. Develop a plan for infrastructure (use of either existing resources or additional capabilities) to support NASA OSS.
> 6. Develop a plan to assess how implementation of OSS moves NASA's Science Mission Directorate (SMD) forward in realizing policy goals, recognizing that it may be a bumpy road and that there will be short-term as well as long-term benefits.
>
> Two or more years prior to OSS requirement:
> 1. Reserve a set portion of software-development funds for OSS.
> 2. Continue incentives and consider competitive funding for replacing priority closed software.
> 3. Continue training and education programs.
> 4. Continue developing infrastructure and management of OSS.
> 5. Continual assessment of implementation against policy goals to refine plans and strategies.
>
> One or more years prior to OSS requirement:
> 1. Reserve a larger portion of software-development funds for OSS.
> 2. Continue training and education programs.
> 3. Continue developing infrastructure and management of OSS.
> 4. Continually assess implementation against policy goals to refine plans and strategies.
>
> Year of full OSS requirement:
> 1. Fund only OSS for new software development (with limited restrictions).
> 2. Continue incentives to open priority-established closed software as appropriate.
> 3. Review the policy annually to ensure that it is aiding NASA SMD's objectives.
> 4. Continue education and maintain infrastructure.
> 5. Continually assess implementation against policy goals to refine plans and strategies.

Box 5.2 describes the general path to take, but different software types will have different priorities and requirements in moving toward openness. Table 5.1 lays out how the various policy options described above could be applied and when and if mandates would be applied. Each of the different software types is then discussed in turn. Of course, there will be variance by discipline and community within each software type as well.

5.4.1 Policy Options Applied to Different Software Types

Table 5.1 lays out how the various policy options described above could be applied and when and if mandates would be applied. Each of the different software types is discussed below.

Libraries

Libraries are products that many people depend on, and the user's ability to verify the software and to adapt it for the user's own purposes is a key feature of software libraries for science. They would be a high priority to encourage to be OSS, and all the incentives are applicable. Mandating that new libraries and projects adding to existing OSS libraries be open source could happen fairly quickly. Existing closed libraries could be incentivized to become open source. Many libraries exist only because of researchers' investments of large amounts of time. Funding key developers and covering expenses will ensure the continuing quality and feature sets of existing libraries.

Single-Use Software

Software written for use in unique instances is inherently lower priority for OSS policy. SMD may want to consider some proposal add-ons or pilots to transition single-use software to more broadly useful OSS, but mandates for opening single-use software would be challenging to implement. There may still be interest in opening the code to reproduce results or broaden a study. Journals may require code used in a publication to be available, because reproducing the results requires the inputs and configurations as well as the software that produced the results. The effort to produce reproducible research papers can be manageable if planned from the outset, but it can be impractical if undertaken after a multiyear project. This cultural shift will require broader incentives and commitment than just NASA can provide and will likely be based on the incentives created in scholarly communication that recognize open practitioners.

Analysis Software

Analysis software may involve a few or many investigators and may have specific or broad applications. It would be a high priority to incentivize openness for newly developed and broadly useful software. Legacy software and software created in collaboration with people not supported by NASA would be a lower priority and harder to mandate because of potential licensing challenges. For older software with poor documentation, testing, and so on, rewriting may be the better solution, if the software has broad use and existing contributors have the first opportunity to rewrite.

Modeling and Simulation Software

Modeling and simulation (M&S) software is in broad use and may receive targeted development funds. The diversity of software in this category is wide, and community cultures span the openness spectrum. Research reproducibility, study replication, and extension of results are all served by opening model software, but priorities for opening different models will vary. Policies in the short term may need to adapt to the research community and model scale (e.g., number of investigators involved). With some large software projects, it may be difficult to locate all the contributors and get them to agree to an OSS license. There may also be additional legal concerns that are specific to individual contributors. For commercial models (e.g., chemical kinetics or spectroscopy) that are in broad use, share-in-savings contracts may be possible to fund by open sourcing them. For many established models, the easiest path to openness may be rewriting the software, which could be financially challenging even if the original contributors are involved.

Frameworks

Modeling frameworks glue together different models. They manipulate model quantities, and it is important for researchers to see exactly how, so they can ensure, for example, that the framework follows relevant conservation laws and does not violate downstream assumptions. Open frameworks are a high priority, and all incentives can be applied to motivate the opening or replacement of key existing frameworks. Mandates may be appropriate for new development while recognizing similar concerns about licensing and documentation, as with M&S software.

Sensor Software

Science requires an understanding of the computations and manipulations made to data. Often, a mission's data center delivers processed data and the processing changes periodically as calibration methods and data improve. It is thus imperative for the software doing these calculations to be available to all investigators, so that investigators can rerun the calculations with parameters and input data of their choosing. For this to happen, the software needs to be open.

All software that manipulates measurements taken by NASA sensors before those measurements are delivered to investigators needs to be open (and ideally, containerized). This may include software running on the spacecraft. As some data-processing centers integrate proprietary database-access software into their calibration process, retroactively applying this recommendation will require negotiating with those centers, and possibly opening only the portions of the software without proprietary storage and retrieval parts. All future contracts for any kind of sensor calibration and data processing software need to specify that the software will be open, documented, and supplied to investigators in a standard form.

Infrastructure Software

Infrastructure software helps manage and store data and helps researchers discover, select, and transfer data. With rapidly growing volumes of data and increasing cloud-based processing, access to the software behind an archive is of increasing interest. Standing up a new data center for a mission is expensive and time consuming. It is in NASA's best interest to open the software in this category. This will enable principal investigator (PI)-run data centers for small missions to provide modern services and will ensure the best competition for larger data center contracts. The variety of existing infrastructure software, and their current states of documentation and dependence on proprietary components, makes it more practical to implement a future policy than a retroactive one. Openness requirements here are thus best handled on a case-by-case basis for existing software (which may be in use for decades), depending on the challenges and benefits involved. This implies that certain software will need to be maintained for a very long time.

5.5 ASSESSMENT AND FUTURE CONSIDERATIONS

As just discussed, there will be a variety of implementation strategies for the different options, disciplines, and software types. An assessment of community software users before, during, and after implementation of policies could help advance policy goals more efficiently. A Web-based survey of almost 2,000 people, 71 percent with a Ph.D. or equivalent, examined how scientists use software and computers for research. The survey found that scientists use computers differently than professional software developers, scientists work more on their own or in small projects, and they are not as familiar with tools relevant to larger projects (e.g., version control, verification, and testing tools).[7] It will be critical for NASA to develop an ongoing assessment program to ensure that OSS implementation is advancing science, fostering collaboration, and generally helping to achieve NASA goals (see Chapter 1). The OMB memo on Federal Source Code Policy directed agencies to "collect additional data concerning new custom software to inform metrics to gauge the performance of this pilot." Any metrics that SMD develops need to be in line with NASA's goals (e.g., not just reuse, publications, citations, etc.). It will also be important to understand how the community is reacting to and adapting to new policies and incentives. SMD may want to consider funding research to explore that community evolution (e.g., Geiger et al. [2018] studies of the Moore-Sloan Foundation Data Science Initiative[8]) as well as conduct workshops, pilot projects, and other mechanisms to directly solicit community feedback. Formal economic analysis of the impact of OSS, similar to the

---

[7] See https://www.americanscientist.org/article/how-do-scientists-really-use-computers.
[8] R.S. Geiger, C. Mazel-Cabasse, C.Y. Cullens, L. Norén, B. Fiore-Gartland, D. Das, and H. Brady, 2018, *Career Paths and Prospects in Academic Data Science: Report of the Moore-Sloan Data Science Environments Survey*, SocArXiv, https://doi.org/10.17605/OSF.IO/XE823; R.S. Geiger, N. Varoquaux, C. Mazel-Cabasse, and C. Holdgraf, 2018, *The Types, Roles, and Practices of Documentation in Data Analytics Open Source Software Libraries*. Computer Supported Cooperative Work (CSCW), pp. 1-36, https://doi.org/10.1007/s10606-018-9333-1.

analyses done for data and described in Chapter 3, could also be helpful. Early assessments are critical in policy implementation and help establish a baseline for future assessments that might strengthen the science justification for broader open science policies.

Policy options will need to continually adapt to changes in software and technology. At present, the field of machine learning, for example, presents new challenges. A recent paper in *Science* studied 400 algorithms presented at two top artificial intelligence (AI) conferences and found that only 6 percent shared software and 33 percent shared data.[9] As stated earlier, releasing data is not sufficient to ensure reproducibility. In machine learning (ML), the way the set is partitioned into training data and testing data may not be documented. Some ML frameworks apportion these subsets automatically and with a degree of randomness ("data shuffling"). The ML community is actively working to address reproducibility concerns (e.g., holding dedicated workshops at the top conferences, such as the International Conference on Machine Learning [ICML] and the Reproducibility Workshop). Research efforts focused on improving ML workflows are currently under way, so this problem may soon be resolved, but NASA will need to assess policy options to require more than source code and input data in order to ensure reproducibility.

> **Conclusion:** SMD will need to assess and adapt policy to new computing technology developments to maintain its intent.

5.6 POLICY IMPLEMENTATION

With policy options for specific software types, NASA SMD also will need to consider implementation details. The committee provided some general recommendations in Chapter 4 that will help establish a general norm of OSS. More specific recommendations on implementing OSS policies, regardless of the options selected for different software types and circumstances, are provided below.

5.6.1 Licensing

Chapter 2 discusses the various legal issues around software release. Legal complexity around software release involves copyright, patents, and legal restrictions on what can be shared. Based on the input received by the committee, understanding of these issues is often lacking, and confusion within the research community and within NASA is common. Applying an appropriate license reduces confusion by clearly indicating the conditions under which software can be reused and redistributed. As defined in Chapter 2, OSS needs to have a standard license (e.g., Open Source Initiative [OSI]-approved), but choosing the right license is still complex. SMD investigators may also contribute code to existing projects, which may bring up issues of license compatibility.

NASA needs to balance the different goals it is trying to achieve with OSS. Industry favors more permissive licenses, whereas some investigators may favor more restrictive licenses or may desire to impose more restrictions until after they publish (see Figure 2.2). License compatibility is a complex yet important issue in ensuring effective software reuse. As noted earlier, the current NOSA license (1.3) is not well respected by the open source community (WP 28).[10,11] Licensing, compatibility, and general legal issues will need to be a major part of the education program recommended in Chapter 4.

> **Conclusion:** NASA will need to balance the goals of enabling innovation, facilitating scientific reproducibility, stimulating the economy, and benefiting society when recommending particular licenses that are as open and permissive as possible and only as restrictive as necessary.

---

[9] See http://www.sciencemag.org/news/2018/02/missing-data-hinder-replication-artificial-intelligence-studies.
[10] *NASA Open Source Summit*, 2011, https://www.nasa.gov/open/source/.
[11] See https://docs.google.com/document/d/1TagS_gwDhDfxjr7WpG78_aIcfoPO1tMXBPeCMEE3-Us/edit?hl=en#.

Recommendation: NASA Science Mission Directorate should encourage the use of standard open source licenses, but not mandate a particular license. Nonstandard licenses should be justified in the software management plan.

Standard licenses include, but are not limited to, those considered popular and widely used by OSI.[12] The use of standard licenses is also recommended by the Federal Source Code Policy's Pilot Program (see Section 3.3.7).

5.6.2 Planning and Facilitating Software Release

Software release is becoming a normal part of the scientific process. The process is often swift, highly collaborative, and exploratory as incomplete solutions are worked out. NASA civil servant scientists need to participate in this process to remain competitive. Currently, software released by NASA employees must undergo a rigorous vetting process before its release, to ensure compliance with software engineering standards (Chapter 3 of NPR 7150)[13] and to prevent disclosure of restricted information. The same process applies for all software regardless of the scale, topic, or a priori risk of the code. The procedure ignores the different software types and can be unnecessarily burdensome at times (e.g., WP 1). The NASA Technology Transfer Program has made major improvements to the process, recently. The program could improve the process further by enabling expedited review for some software.

Certain software requires rigorous and thorough review (e.g., software near export control boundaries or that includes commercial-off-the-shelf products), but much of the software developed by NASA SMD does not. Creating a streamlined review process for low-risk software would enhance the ability of NASA scientists to work more openly with the rest of the academic community and provide a model for other institutions. For example, Elsevier normally approves OSS release normally within 2 weeks (WP 30).

Expedited review could be based on current definitions of confidential information and the risk that a work could develop new knowledge in areas of security concern. For the vast majority of civil-servant authors, software review could be simple and quick, saving time and money without adding national security or legal risk.

For software in sensitive areas, identifying the likely risks in advance and working with legal experts in planning the software could reduce risk, while expediting the final review by focusing it on the areas of concern. For example, risky code might sometimes be limited to a single module that could be replaced or omitted in public release, while the full code may be made available under appropriate safeguards to trusted citizens.

Conclusion: NASA's current, internal software release policy can cause undue and potentially harmful delay in the release of low-risk software.

Universities and other research organizations also have policies to control software release. Often these policies are designed to protect intellectual property rights of the institution. Investigators can usually release their software more easily if the open source arrangements were made prior to award of a grant or contract.

Recommendation: NASA Science Mission Directorate should develop internal policies and external legal language conducive to the swift release of open source scientific software, and the full participation of NASA employees in internal and external open source projects, without jeopardizing national security or incurring legal liability.

To implement this recommendation, SMD will need to work in coordination with NASA Technology Transfer Office and export control offices.

---

[12] See https://opensource.org/licenses.
[13] See https://nodis3.gsfc.nasa.gov/displayDir.cfm?t=NPR&c=7150&s=2B.

5.6.3 Ongoing Compliance

Any OSS policy based purely on licensing considerations could possibly be circumvented. For example, software as a service (SaaS)—a delivery model where users access software and data through a Web interface, like at the CCMC—avoids distributing the code and sidesteps license requirements. Services can be free or require payment. Examples could be as simple as a free Web interface that collocates user-given data points with contemporaneous satellite data and as complex as a paid subscription to high-resolution ocean wave forecasts for shipping. Since the software itself is not copied, copyright license terms may not be triggered. Recent versions of the GNU General Public License (GPL) do require disclosure of software deriving from GPL-licensed work when used in a SaaS environment, but permissive OSS licenses do not. This raises concerns about source code access, reproducibility, and long-term sustainability and maintenance. Some third-party vendors provide useful SaaS, such as machine learning systems and software implementing patented algorithms. Reproducibility of the science may be impacted by SaaS systems, especially generic ones that may change without notice to the user, or that may disappear entirely during an investigation. Software management plans will need to include discussion of use of SaaS in new software development.

Conclusion: SaaS and other computing technologies can be used in a positive way but may also be used as a mechanism to circumvent policy.

6

Discussion

Software increasingly underpins science. It connects the data, hardware, networks, and people to enable the analysis leading to new knowledge and discovery. Open source software (OSS) can be a major facilitator of this process by helping create an environment for collaboration and transparency. This can provide efficiencies, aid in scientific reproducibility, and help NASA advance its mission. Movement toward more OSS development will foster a norm of openness, collaboration, and sharing, but it is not a straightforward process. NASA will need to balance the different goals in Chapter 1 of enabling innovation and discovery, facilitating scientific reproducibility, encouraging collaboration, ensuring security, and benefiting society when developing policies that are as open as possible and only as closed as necessary.

The basic act of releasing software as open source is not difficult (see Figure 4.2), but it can evoke some complex considerations, including complying with institutional and contractual intellectual property guidance, determining the appropriate license to apply, and ensuring that export-controlled information is not included in the software. Furthermore, making software open is insufficient by itself in ensuring scientific reproducibility or realizing any of NASA's goals. Poorly documented software and associated data files, even if shared publicly, will likely result in an inability to replicate research. Historically, software funding and support has not been well coordinated across NASA's Science Mission Directorate (SMD). Understandably, the focus is often on creating new software to solve a scientific problem rather than evolution, maintenance, and sharing of existing software. Yet, to make open software truly useful requires a coordinated, end-to-end development approach supported by adequate infrastructure, community practices, and education over the long term. Any open source policy will need to address these issues. Open source can provide great benefits, but there will be transition and maintenance costs requiring a careful balance of trade-offs and active engagement by program managers. The committee's nine recommendations regarding an OSS policy are restated below.

Recommendation: NASA Science Mission Directorate should explicitly recognize the scientific value of open source software and incentivize its development and support, with the goal that open source science software becomes routine scientific practice.

Recommendation: NASA Science Mission Directorate should initiate and sponsor programs to educate and train researchers in open source best practices. Topics could include, but are not limited to export controls, licensing and intellectual property, workflows, and software development. These resources

could be made available to the community via in-person trainings as well as webpages, screencasts, and webinars.

Recommendation: Any open source software policy that NASA Science Mission Directorate develops should not impose an undue burden on researchers; therefore, any policy should be as simple as possible, and any mandates should be fully funded.

Recommendation: NASA Science Mission Directorate should support the infrastructure, governance, and maintenance of a healthy open source community, taking advantage of existing community resources to the greatest extent possible.

Recommendation: NASA Science Mission Directorate should support open source community-developed libraries that advance NASA science.

Recommendation: NASA Science Mission Directorate should foster career credit for scientific software development by encouraging publications, citations, and other recognition of software created as part of NASA-funded research.

Recommendation: NASA Science Mission Directorate should consider a variety of policy options depending on discipline and software type and transition to greater openness over time.

Recommendation: NASA Science Mission Directorate should encourage the use of standard open source licenses, but not mandate a particular license. Nonstandard licenses should be justified in the software management plan.

Recommendation: NASA Science Mission Directorate should develop internal policies and external legal language conducive to the swift release of open source scientific software and the full participation of NASA employees in internal and external open source projects, without jeopardizing national security or incurring legal liability.

It is important to note that most of these recommendations apply regardless of whether NASA SMD explicitly requires OSS. The cultural shift toward greater openness is much more challenging than the actual policy development. Implementation of any new policy can result in community resistance, where scientists apply to other agencies for funding, attempt to circumvent the spirit of the policy, or ask for special exemptions. This can result in decreased productivity and costly missteps. There are numerous studies in behavioral science on how organizations can best implement change and companies that specialize in consulting on these issues. There are several key elements for reducing the cost and risk associated with implementing new policies, which the committee's recommendations address, as follows:

1. Communicate goals to the community and emphasize the benefits.
2. Identify and empower influential individuals as groundbreakers.
3. Anticipate and try to mitigate obstacles.
4. Provide education, tools, and training to ease adoption.
5. Highlight successes.
6. Implement change incrementally and respond and adapt to feedback.

There are two communities that need to be considered in this process: the program managers who will implement the policy and the research community the policies will impact. Engaging and educating program managers on OSS development best practices will allow them to better anticipate possible obstacles and provide immediate feedback to NASA on how best to affect the culture within their research community. A SMD-level coordination of program managers' experiences as policy is implemented may significantly decrease the length of any transition period. Many of the lessons learned from the implementation of open data policies are discussed in Chapter 3 and can be applied to the implementation of an OSS policy. A theme of this report, however is that OSS policy is more complex than open data policy and will require a multidimensional approach as outlined in the options discussed in Chapter 5. The committee believes in the benefit of an OSS policy, but its implementation must be carefully planned, gradually implemented, appropriately funded, and well coordinated across SMD.

Appendixes

A

Statement of Task

The National Academies of Sciences, Engineering, and Medicine will establish an ad hoc committee to investigate and recommend best practices for NASA as it considers whether to establish an open code and open models policy, complementary to its current open data policy. In carrying out the study the committee will:

- Review and describe examples of code/modeling policies developed by research teams and communities in the NASA-supported disciplines of Earth Science and Applications from Space, the Space Sciences, and other research communities, as appropriate;
- Develop a set of lessons learned from these established approaches paying particular attention to issues such as, but not limited to, proprietary, export control, code/model maintenance, and documentation considerations;
- Define and describe options for policies on open codes and open models for research supported by NASA Science Mission Directorate (SMD) and assess the pros and cons of these options from the perspective of the research community and the interests of NASA; and
- Recommend a set of best practices for NASA to consider should SMD decide to adopt an open code/open model policy for research supported by the agency. The committee may also choose to present alternate sets of best practices rather than just one recommended set.

B

Copyright Issues of Interest to NASA Investigators and Developers of Software

Copyright law is complicated. NASA investigators and software developers would benefit from having some basic awareness of and training in how copyright applies to them.[1] This appendix is not a comprehensive overview of the law, but instead presents some of the key aspects not already covered in Section 2.4. As experience levels in this realm vary, the information is presented in question-and-answer format.

What is copyright and how it is it obtained?

Copyright law grants creators a bundle of exclusive rights that allow the owner of a creative work to prohibit others from copying, distributing, performing, adapting, or otherwise using the work in violation of those exclusive rights. Copyright comes into existence the moment an original work is created. Registration of a copyright is not necessary, and there is no need to include a copyright notice on a work. Creators automatically own an "all rights reserved" copyright at that moment whether they want it or not.

What is the purpose of copyright?

There are two primary rationales offered for copyright. One view is that copyright is designed to provide an incentive to create new works. In many cases, individuals will not invest in the creation of new works without knowing that the work can be exploited by them only for a period of time. Another view, more dominant outside the United States although it is offered by some scholars in the United States, is that copyright ensures attribution for authors and preserves the integrity of creative works.

How is copyright enforced?

The owner of a copyrighted work can bring a lawsuit against any person that violates one of the exclusive rights that copyright grants them. Unauthorized use constitutes infringement and gives the holder of the copyright the right to recover damages. If the work has been registered with the U.S. Copyright Office, the owner is entitled to either actual damages and profits or statutory damages, injunctive relief, and attorney's fees, among

---

[1] A. Morin, J. Urban, and P. Sliz, 2012, A quick guide to software licensing for the scientist-programmer, *PLoS Computational Biology* 18(7):e1002598, https://doi.org/10.1371/journal.pcbi.1002598.

other remedies. Damages can be extremely high depending on the number of infringing copies made, if statutory damages are sought in lieu of actual damages, and if the infringement is willful or innocent.

What is fair use and does it protect scientists and researchers?

Copyright law has built-in limitations designed to ensure that the rights of the public are not unduly burdened by copyright. One of those limitations in the United States is *fair use*. Fair use is designed to promote the creation and sharing of knowledge. It is a defense to a claim of infringement, allowing the public to do certain things with a copyrightable work without infringing the rights of the owner. Fair uses can include using a work in commentary, research, and scholarship. Fair use is a highly factual analysis—there are no fixed and certain categories of uses on which a user can always depend. Courts look at the particular facts before making the determination. For example, not every use of another person's copyrightable content in a research paper is a fair use.

Fair use applies to a particular user and his or her use of the work; it does not automatically apply to reuse of the work by subsequent users. For example, if a scientist includes a colleague's copyrightable chart in slides presented at an educational conference, the scientist's use may be a fair use. But if the slides are then posted on the Internet and downloaded and sold by a researcher, the researcher cannot depend on the fair use made by the scientist for that particular use and may be liable for infringement.

What is the public domain?

Generally, the *public domain* is defined as consisting of works and other materials that are not protected by copyright for any of the reasons described in Section 2.4. Anyone can use a work in the public domain for any purpose, including commercially. As a matter of copyright law (although not necessarily scientific and scholarly norms), users of public domain materials are not required to attribute or credit the creator, although falsely claiming to have copyright in a public domain work may constitute copyfraud. Derivative works can be made of public domain works, although the copyright in the derivative work extends only to the new original elements that have been added.

Copyright terms vary by country. While the general term of copyright in the United States for works created by individuals lasts for the life of the creator plus 70 years, in much of the world the term is life plus 50 years, and in a few countries life plus 100 years. Because copyright terms differ, a work in the public domain in the United States may be not be in the public domain everywhere.

What is the role of standardized licenses in advancing research and science, including software development?

Standard licenses and the routine release of copyright interest to software contributions are critical to the success of community-developed software, often including scientific software. If these are not organized in advance, it can be essentially impossible to unencumber the work later. In some cases, institutional intellectual property offices are focused on exploiting work for profit, and they are slow to agree to the open sharing on which science depends.

What related career risks are there?

Most scientists work for employers who administer funds (such as NASA grants) to them. *The work-for-hire clause puts the ownership of essentially all their creative work into the employer's hands, not the scientist's.* If a scientist moves to another organization, the original organization has the right to take full possession of everything the employee created, and even to give it to another employee for further exploitation, or to sell it. The sponsor (e.g., NASA) has limited rights in such cases. In some organizations, employees have negotiated the right to keep ownership, or to have it transferred to follow them wherever they go, but this is not universal. It is important for all scientists to know the ownership status of their work, and it is in their best interest to negotiate favorable terms upon hire. Some NASA-funded scientists have lost access to their life's work because they left an employer who retained ownership of the work.

C

Call for White Papers and Listing of Received White Papers

CALL FOR WHITE PAPERS

Dear Colleagues,

NASA has requested the National Academies of Sciences, Engineering, and Medicine to investigate and recommend best practices for NASA as the Science Mission Directorate considers whether to establish an open code policy, complementary to its current open data policy. The committee appointed by the Academies to carry out this study is now soliciting community input in the form of white papers. Full details of the committee's membership and schedule of activities, as well as the statement of task for this study, are available at http://sites.nationalacademies.org/SSB/CurrentProjects/SSB_178892.

The specific goal of this call for white papers is to hear broadly from the community on any issues, situations, or points of view relevant to the topic, to ensure consideration of the full set of possible consequences of any new NASA open source policy. For the purpose of this call, "open code" and "open source" are synonymous and refer to computer program source codes released publicly under an open source license, as defined by the Open Source Initiative, https://opensource.org/licenses.

To be considered at the committee's next meeting, white paper submissions are due no later than January 12, 2018. The committee strongly encourages authors to submit white papers by this deadline, but papers will continue to be received until January 31, 2018.

As a guide, the committee suggests the following topics for consideration:

1. What positive and negative impacts would arise for you, your workplace, your NASA-funded research, science in general, education, commerce, society, and so on, if all future NASA-funded science code were required to be open source? For example, what maintenance and support issues might arise from open source policies that would not otherwise arise? What relevant experiences have you had with science codes owing to sharing or access constraints? How might negative impacts be mitigated?
2. What would be the consequences, positive or negative, if NASA exercised any rights it may have to require that existing codes previously developed under NASA funding be made open source?
3. If a future policy is in place that would require all NASA-funded science codes to be made available under an open source license, what exceptions, if any, might be made to this policy? What principles might be applied in

granting and then overseeing such exceptions, and what parallel measures could be taken to mitigate any detrimental effects an exception might have on code availability and re-use?
4. What lessons can be drawn from your experience with open data policies that might help inform future open source policies?
5. What policy differences, if any, might be considered for NASA-funded science codes produced as part of a research grant versus those produced under other NASA funding mechanisms, such as contracts, interagency transfers, or cooperative agreements? Might there be different policy requirements for various types of code (such as models, libraries, modules, etc.) or codes produced by various types of research groups (for example, individuals or modeling centers)?
6. What special (non-obvious) considerations might exist for codes with multiple funding sources or codes that incorporate proprietary libraries or other restricted information, such as International Traffic in Arms Regulations (ITAR)-regulated code?
7. What non-policy approaches could NASA take to encourage open source licenses for NASA-funded codes (for example, bounties for opening closed codes or for creating new open codes that do the same tasks as closed codes; badges on published papers indicating open source, open data, and reproducible-research; mechanisms for giving career credit for compliant research products like these)? How might these approaches be implemented and what potential issues could be envisioned regarding enforcement of these kinds of practices?
8. Over the long run, what would be the impact on the quality and reproducibility of research if NASA required all NASA-funded, peer-reviewed science papers to include an electronic compendium of (or pointers to) the source codes, inputs, and outputs that produced each scientific claim in the paper?
9. Other issues you would like the committee to consider.

Please note that the suggested topics and questions, as well as the manner in which they are framed above, should not be seen as a preview of any findings or recommendations the committee may make.

Guidelines for White Paper Format and Submission

If you have an opinion on any relevant matter, please submit a white paper, following these guidelines:

1. The suggested topics are broad areas intended to initiate thought. A paper need not (and generally should not) address all questions in a given list item, above.
2. White papers may not exceed 5 pages in length. This includes all figures, tables, references, and appendices. Web links to other documents may be included in the references.
3. Documents should be single spaced, use 12-pt font, and have 1-inch margins on all sides.
4. Only papers submitted through the online submission process will be accepted. Required entries are title (max. 150 characters), short summary (max. 350 characters), authors, corresponding author email address and telephone number.
5. Only papers in Microsoft Word (.doc, .docx) and Adobe Acrobat (.pdf) formats will be accepted.
6. A cover page may be included and will not count toward the 5-page limit. It should state the title of the white paper, the primary author's name, phone number, institution, and email address. All authors who contributed significantly to the text must be named on the cover page, including: full name, position, affiliation, and how they are a stakeholder. The permission of each co-author must be explicitly given prior to submission.
7. Appendices may contain license or policy examples or other supporting, pre-existing documents, but not further text or other material created for the paper.
8. Contributions are public and fully attributed (i.e., not anonymous). If not already in the public domain, copyright release is required at time of submission.
9. Group submissions are strongly encouraged. We encourage community discussion to consolidate similar papers. A numbered list of supporters who did not contribute significantly to the text may be attached as the first appendix. Supporters listed must include the same information as for authors on the cover page.

Please respect that the committee is not large and has a short time to evaluate a potentially large number of white papers. A well argued, concise paper will make the strongest impression. Use specific examples from your own experience, cite specific policies that impact you, use numbers, etc., wherever possible. *When it is not obvious*, relate the argument to NASA Science Mission Directorate's goals.

LISTING OF SUBMITTED WHITE PAPERS

The white papers listed in Table C.1 are numbered generally in order of submission, and some numbering irregularities arose from duplicate or replacement submissions.

TABLE C.1 White Papers Submitted to the Committee

Ref. #	Title	Submitter
1	Open Source Code, from the Perspective of a Scientist at a NASA Center	Jane Rigby, NASA Goddard Space Flight Center (GSFC)
2	Earth Science Data Systems: Policy for Open Source Software Governance	Chris Mattmann, NASA Jet Propulsion Laboratory (JPL)
3	The NASA-Funded EPIC Atmospheric Model: Advantages of Open-Code Status since 1998	Timothy E. Dowling, University of Louisville
4	Book Performance Report: 2016 [for "The Dawn Mission to Minor Planets 4 Vesta and 1 Ceres"]	Christopher Russell, University of California, Los Angeles
5	White Paper in Support of NASA's Proposed Open Code Data Policy	James Paul Mason, NASA GSFC
6	Comments on a Future Open Code Policy: Potential Problems and Pitfalls	Daniel Weimer, Virginia Tech
7	Best Practices for a Future Open Code Policy for NASA Space Science: Response to a Call for White Papers	Peter Young, George Mason University
8	Open Source White Paper	Brian R. Dennis, Joel Allred, Charles N. Arge, Gordon D. Holman, Andrew Inglis, Richard Schwartz, Albert Shih, Anne K. Tolbert, and Dominic Zarro, NASA GSFC
9	Software Engineers' Perspective on Open Source Projects at NASA/GSFC	Chiu Wiegand, Rick Mullinix, and Justin Boblitt, NASA GSFC
10	Practical Considerations of Open Source Delivery	Eric Lyness, Microtel, LLC
11	White Paper on Possible NASA SMD Open Code Policy and Practices	Charles H. Acton, JPL
12	Contract Language and Software Redistribution at NASA	James Vasile and Karl Fogel, Open Tech Strategies
13	In Support of an Open Code Policy Which Is Inclusive of Commercial Technologies to Accelerate Reproducibility of Science	Tripp Corbett, Dawn Wright, and Marten Hogeweg, Esri
14	Open Source to Serve Community Science	Arfon Smith, Kenneth Sembach, Nancy Levenson, Thomas M. Brown, Marc Postman, Neill Reid, Massimo Stiavelli, and Roeland van der Marel, Space Telescope Science Institute
15	Open Source Code and Intellectual Property	Stanley C. Solomon, National Center for Atmospheric Research
16	Perspectives on Reproducibility and Sustainability of Open-Source Scientific Software from Seven Years of the Dedalus Project	Jeffrey S. Oishi, Bates College; Benjamin P. Brown, University of Colorado, Boulder; Keaton J. Burns, Massachusetts Institute of Technology; Daniel Lecoanet, Princeton University; Geoffrey M. Vasil, University of Sydney
17	A Recommendation for a Complete Open Source Policy	Steven D. Christe, NASA GSFC; Jack Ireland, ADNET Systems, Inc.; Daniel Ryan, NASA GSFC
18	Comments for Open Code Policy for NASA SMD	V.G. Merkin, K. Sorathia, L. Daldorff, A. Ukhorskiy, and M. Sitnov, JHU/Applied Physics Laboratory; J. Lyon, Dartmouth College

APPENDIX C

TABLE C.1 Continued

Ref. #	Title	Submitter
19	The Role of Commercial Software in an Open Source World	Zachary Norman and Daniel Platt, Harris Geospatial Solutions, Inc.
20	[No Title Given]	Robert E. Grimm, Southwest Research Institute
21	What Does Scientific Reproducibility and Productivity Really Mean? The Dangers and Difficulties of a Blanket Open Code Policy	John T. Emmet, Naval Research Laboratory (NRL); Jens Oberheide, Clemson University; Douglas P. Drob, McArthur Jones, Jr., Fabrizio Sassi, David E. Siskind, and Kate A. Zawdie, NRL
22	Implications of a Future NASA SMD Open-Source Policy	C. Richard DeVore, Spiro K. Antiochos, Alex Glocer, Judith T. Karpen, James E. Leake, and Peter J. MacNeice, NASA GSFC
23	Software Practices for Improved Collaboration among Space Scientists	Asti Bhatt, SRI International; Ryan McGranaghan, NASA JPL; Tomoko Matsuo, University of Colorado; Yolanda Gil, University of Southern California
24	Towards Reproducibility Using Open Development: Astropy as a Case Study	Erik Tollerud, Space Telescope Science Institute
25	An Open Source Approach for NASA	Anthony J. Mannucci, Olga Verkhoglyadova, Ryan McGranaghan, Giorgio Savastano, and Bruce Tsurutani, NASA JPL
27	Impacts, Consequences, and Perspectives on a Future Open Code Policy for NASA Space Sciences	Ross A. Beyer, NASA Ames Research Center
28	No to NOSA, Yes to Mainstream Licenses	Ross A. Beyer, NASA Ames Research Center; Terry Fong, NASA STMD; Mark B. Allan, Stinger Ghaffarian Technologies, Inc.; Jason Laura and Moses P. Milazzo, U.S. Geological Survey; Robert G. Deen and Wayne Moses Burke, NASA JPL
29	Our View on Open Source Code Cevelopment for Scientific Software at the Center for Space Environment Modeling at the University of Michigan	Gabor Toth, University of Michigan
30	Answers to Committee Questions	Brad Fenwick, Elsevier
31	AFRL Response	Cheryl Huang, Air Force Research Laboratory
32	White Paper on Release Requirements for Legacy Model Codes	Mark Marley, NASA Ames Research Center; Jonathan Fortney, University of California, Santa Cruz; Richard Freedman, SETI Institute; Peter Gao, University of California, Berkeley; Roxana Lupu, BAERI; Caroline Morley, Harvard University; Tyler Robinson, Northern Arizona University; Didier Saumon, Los Alamos National Laboratory
33	Open Source Software as the Default for Federally Funded Software	Travis E. Oliphant, Quansight, LLC
34	NASA Science Centers Need to Support and Lead Open Source Development or Become Obsolete	Tess Jaffe, T. Barclay, and P. Boyd, NASA GSFC
35	Comments on Best Practices for a Future Open Code Policy for NASA Space Science	J.D. Huba, NRL
37	Space Weather Prediction Center Support of NASA Open Code Policy	Steven M. Hill, Eric Adamson, Michele Cash, Marcus England, and Joe Schoonover, NOAA Space Weather Prediction Center
38	Current and Future considerations for a NASA Open-Code Policy	Adam Kellerman, Steve Morley, and Alexa Halford
39	Assuring Positive Value for Open-Source Software	Thomas J. Loredo, Cornell University
40	Reproducible Science via Open Source Requirements: Increasing Impacts of and Public Support for NASA Mission Science	Michael Hirsch, Boston University

continued

TABLE C.1 Continued

Ref. #	Title	Submitter
41	Open Code Policy for NASA Space Science: A Perspective from NASA-Dupported Ocean Modeling and Ocean Data Analysis	Sarah Gille, Scripps Institution of Oceanography, University of California, San Diego (UCSD); Ryan Abernathey, Lamont-Doherty Earth Observatory, Columbia University; Teresa Chereskin, Scripps Institution of Oceanography, UCSD; Bruce Cornuelle, Scripps Institution of Oceanography, UCSD; Patrick Heimbach, University of Texas, Austin; Matthew Mazloff, Scripps Institution of Oceanography, UCSD; Cesar Rocha, Scripps Institution of Oceanography, UCSD; Saulo Soares, University of Hawaii; Maike Sonnewald, Massachusetts Institute of Technology; Bia Villas Boas, Scripps Institution of Oceanography, UCSD; Jinbo Wang, JPL
42	Considerations for a Future Open Code Policy for NASA Space Science	Dana Akhmetova, KTH Royal Institute of Technology, Sweden; Jan Deca, University of Colorado, Boulder
43	Considerations Regarding the Proposed Open Code Policy	Cody Wiggs, University of Colorado, Boulder
44	Best Practices for a Future Open Code Policy: Experiences and Vision of the Astrophysics Source Code Library	Lior Shamir, Lawrence Technological University; Bruce Berriman, Caltech/IPAC-NExScI; Peter Teuben, University of Maryland; Robert Nemiroff, Michigan Technological University; Alice Allen, Astrophysics Source Code Library

D

Biographies of Committee Members and Staff

COMMITTEE

CHELLE L. GENTEMANN, *Co-Chair,* is a senior scientist at Earth and Space Research, where she works on remote sensing, air-sea interactions, upper ocean dynamics, and sea surface temperatures (SSTs). Prior to that, Dr. Gentemann was with Remote Sensing Systems, where she focused on air-sea interactions, diurnal warming, passive-microwave SST retrievals, instrument calibration, and radio frequency (RF) interference. Dr. Gentemann participates in a number of science teams and committees, including the Group for High Resolution Sea Surface Temperatures (GHRSST). She has been the principal investigator (PI) for the U.S. component of GHRSST, the Multi-sensor Improved Sea Surface Temperature project, since 2003. She was awarded the National Oceanographic Partnership Program's Excellence in Partnering Award and the American Geophysical Union (AGU) Charles S. Falkenberg Award. She received her Ph.D. in meteorology and physical oceanography from the University of Miami. Dr. Gentemann has served on several committees of the National Academies of Sciences, Engineering, and Medicine, including the Committee on Earth Science and Applications from Space and the Committee on a Framework for Analyzing the Needs for Continuity of NASA-Sustained Remote Sensing Observations of the Earth from Space.

MARK A. PARSONS, *Co-Chair,* is a senior research scientist at Rensselaer Polytechnic Institute (RPI). Mr. Parsons is also director of data science operations for the Tetherless World Constellation at RPI. Previously, he was secretary general of the Research Data Alliance and an associate director of the Rensselaer Institute for Data Exploration and Applications. Prior to that, he was lead project manager at the National Snow and Ice Data Center at the University of Colorado, Boulder. He has been involved in data management for more than 20 years, during which he defined and implemented comprehensive data management processes for many projects and organizations. He is active in multiple international informatics efforts and led the data management effort for the International Polar Year (IPY). Mr. Parsons is a member of the Foundation for Earth Science Information Partners board of directors and a member of the Coordinating Committee for the Transparency and Openness Promotion (TOP) Guidelines. He received the AGU/Earth Science Information Partners Charles S. Falkenberg Award. He earned his M.A. in geography at the University of Colorado, Boulder. Mr. Parsons has served on the Committee on the Development of a Strategic Vision and Implementation Plan for the U.S. Antarctic Program, as an ex officio member of the Board on Research Data and Information, and as a member of the Committee on Archiving and Accessing Environmental and Geospatial Data at NOAA.

LORENA A. BARBA is an associate professor of mechanical and aerospace engineering at George Washington University. Dr. Barba's research interests include computational fluid dynamics, high-performance computing, computational biophysics, and animal flight. She was an early adopter of GPU technology for scientific computing. She has advocated for open source software for science and open educational resources for years, and her research group is well known for its open science practices. Dr. Barba is a member of the board of directors for NumFOCUS, a 501(c)(3) nonprofit that supports and promotes world-class, innovative, open source scientific computing. She is a member of the editorial board for IEEE/AIP Computing in Science and Engineering (leading a new track on Reproducible Research), *Journal of Open Source Software* (founding member), and *ReScience Journal*. Dr. Barba received the National Science Foundation (NSF) Faculty Early CAREER award and was named a CUDA fellow by NVIDIA Corporation in 2012. She is an awardee of the U.K. Engineering and Physical Sciences Research Council (EPSRC) First Grant program, is an Amelia Earhart Fellow of the Zonta Foundation, and was awarded a Leamer-Rosenthal Prize by the Berkeley Institute for Transparency in the Social Sciences (BITSS) in the Leaders in Education category. Dr. Barba earned her Ph.D. in aeronautics from the California Institute of Technology.

KELLE L. CRUZ is an associate professor at the City University of New York Hunter College in the Department of Physics and Astronomy. Dr. Cruz is also a research associate in the Astrophysics Department at the American Museum of Natural History. Her research interests include the study of low-mass stars and brown dwarfs using optical and near-infrared spectroscopy. She is a member of the coordinating committee of the Astropy Project and is also the founder and editor of the AstroBetter Blog and Wiki. Previously, Dr. Cruz was an NSF Astronomy and Astrophysics postdoctoral fellow at the American Museum of Natural History and a Spitzer Postdoctoral fellow at Caltech. She is currently a councilor/trustee of the American Astronomical Society (AAS) and served previously as the chair of the Employment Committee. Dr. Cruz earned her Ph.D. in physics and astronomy from the University of Pennsylvania.

BRENDA J. DIETRICH is an IBM fellow and vice president at IBM Business Solutions. As a fellow, author, inventor, and leader in analytics and data science, Dr. Dietrich applies data and computation to processes throughout IBM and IBM clients. She led the Mathematical Sciences Department in IBM Research for over a decade. She was IBM's chief technology officer for Business Analytics, led emerging technologies in Watson, established Data Science for Insight Cloud Services, and is currently leading data science activities in The Weather Company, a newly acquired IBM Business. Dr. Dietrich's research interests include mathematical models of decision processes, particularly those related to the allocation of resources; use of data and computation in decision making, both in enterprise processes and in individual choices; use of computational methods such as visualization, statistics, data mining, simulation, and optimization to generate and evaluate decisions; extraction of models that describe the operation of systems, both physical and behavioral, from data, especially data generated by automation of business processes and computer intermediation of social processes; and cognitive computing and extending the base capability of natural language processing and search-based methods to include structured data analysis and interpretation. Dr. Dietrich received a B.S. in mathematics from the University of North Carolina, Chapel Hill, and an M.S. and a Ph.D. in operations research from Cornell University. She has served on the National Academies Industrial, Manufacturing, and Operational Systems Engineering Peer Committee, the 2019 Nominating Committee, and the Panel on Assessment and Analysis at the U.S. Army Research Laboratory (ARL).

CHRISTOPHER L. FRYER is a Scientist 5 at Los Alamos National Laboratory (LANL) in the Computer Science, Computational Science, and Statistics Division. At LANL, Dr. Fryer is the director of the Center for Theoretical Astrophysics and the project lead of the high-energy density physics impact team. Dr. Fryer's research includes a broad range of astrophysical transients (supernovae, gamma-ray bursts, etc.), neutron star and black hole systems, and nucleosynthesis. He also works on laboratory physics experiments at the National Laboratories and has worked extensively on code development and support. At LANL, he is on the advisory committee for the Center of Non-Linear Studies, the Center for Space and Earth Science, and the Information Science and Technology Institute. He also is on the LANL Nuclear Particle Astrophysics and Cosmology Senior Review Team and the Board for Institutional Computing. For his work on multidimensional simulations of core-collapse supernovae, he was named an APS fellow, and for this work, combined with his laboratory physics work, he received the E.O. Lawrence Award and was named a LANL fellow. Dr. Fryer earned his Ph.D. in astronomy from the University of Arizona.

JOE GIACALONE is a professor of planetary sciences at the University of Arizona in the Lunar and Planetary Laboratory. Dr. Giacalone's research focus is on the origin and physical processes involved in creating high-energy charged particles from near the sun, throughout the heliosphere, and beyond, and how these high-energy particles move throughout the solar system. He has been directly involved with a number of NASA spacecraft missions, including Ulysses, ACE, and Voyager, and is currently a co-investigator for the upcoming Parker Solar Probe mission. Dr. Giacalone uses a wide array of theoretical and computer modeling techniques in his research, including cosmic-ray transport, particle-in-cell kinetic, and magneto-hydrodynamic fluid simulations. Previously, he was a senior research associate at the University of Arizona, and a postdoctoral research associate at Queen Mary, University of London. He is a recipient of an Early Career Award from the NSF and the Professor Leon and Pauline Blitzer Award for Excellence in Teaching of Physics and Related Science at the University of Arizona. He earned his Ph.D. in physics from the University of Kansas. He has served on the National Academies Panel on Physics and on the Panel on Solar and Heliospheric Physics.

SARA J. GRAVES is the director of the Information Technology and Systems Center, a Board of Trustees University Professor, and professor of computer science at the University of Alabama, Huntsville. Dr. Graves directs research and development in sustainable distributed data infrastructures, data mining and knowledge discovery, semantic technologies, information analytics, and cyber security/resilience. Dr. Graves is a member of the Gulf of Mexico Coastal Ocean Observing Systems (GCOOS) board of directors, part of the Integrated Ocean Observing System. GCOOS seeks to facilitate the establishment of a sustained and integrated observing system for the Gulf of Mexico. Dr. Graves is currently a member of the Southeastern Universities Research Association board of trustees and was a founding member of the National Oceanic and Atmospheric Administration (NOAA) Science Advisory Board Data Archives and Access Requirements Working Group and the Climate Change Science Institute Science Advisory Board of the Department of Energy (DOE) Oak Ridge National Laboratory. She has also served as a member of the NASA Headquarters Earth System Science and Applications Advisory Committee and as chair of the ESSAAC Subcommittee on Information Systems and Services. Dr. Graves has been the PI on many research projects with NASA, NOAA, NSF, DOE, and the Department of Defense (DOD). She received her Ph.D. in computer science from the University of Alabama, Huntsville. She has served on the Gulf Research Program Advisory Board and the Board on Research Data and Information.

JOSEPH HARRINGTON is a professor of planetary science in the Department of Physics at the University of Central Florida. Dr. Harrington co-founded the physics Ph.D. track in planetary sciences and is leading its transition into an independent Ph.D. program. Dr. Harrington leads the Spitzer Exoplanet Target of Opportunity Program, an international collaboration of planet hunters and specialists in low-signal data analysis. The group has used NASA's Spitzer Space Telescope to make the first measurements of dozens of exoplanetary atmospheres, including numerous high-impact results, by developing state-of-the-art methods for removing systematics from Spitzer data. As part of this effort, Dr. Harrington led teams that developed several open source scientific codes, including Bayesian Atmospheric Radiative Transfer, which retrieves atmospheric parameters from exoplanetary eclipse and transit data. He wrote the Reproducible Research Software License to prompt a discussion on the robustness of research results involving complex computer codes in astrophysics and beyond. He is the lead organizer of the ExoClimes workshop series and founded the NumPy Documentation Project, which crowd-sourced the documentation of a nascent, now popular, open source numerical programming package. Prior work includes study of cometary impacts into giant planets, the detection of atmospheric waves in Jupiter's atmosphere, and stellar occultations by Saturn's atmosphere and rings. Previously, Dr. Harrington was a researcher at Cornell University and a National Research Council fellow at NASA's Goddard Space Flight Center (GSFC). He earned his Ph.D. in planetary science from the Massachusetts Institute of Technology.

ELVA J. JONES is a professor and chair of computer science of the Department of Computer Science at Winston-Salem State University. Dr. Jones is engaged in study of space science information systems, assistive robotics, computer science education, and assessment methods. Her research interests include artificial intelligence, robotics, computer science education, assessment, game development, gamification, information retrieval, and systems design for decision support. She is the recipient of the Fifty Most Important African Americans in Technology Award; Information Technology Senior Management Forum Ivory Dome Education Leadership Award; Scott Cares Foundation Humanitarian Award for Achievements in Technology; Phi Beta Sigma Outstanding Educator Award; WSSU Sponsored Programs "Million Dollar" Award; City of Winston-Salem Outstanding Women Leaders Award; NASA University Joint Venture (JOVE) Research Award; NASA JOVE Curriculum Development Award; and the NASA JOVE Fellow at NASA GSFC. She is a member of the North Carolina Space Grant executive committee and previously served as a commissioner for the Accreditation Board for Engineering and Technology. She is a member of the Institute of Electrical and Electronics Engineers (IEEE) and the Association for Computing Machinery. Dr. Jones earned her Ph.D. in industrial and systems engineering with a focus in computer studies at North Carolina State University, Raleigh.

MARIA M. KUZNETSOVA is an astrophysicist in the Space Weather Laboratory and the director of the Community Coordinated Modeling Center (CCMC) at NASA GSFC. Dr. Kuznetsova's research interests include global MHD modeling of magnetosphere dynamics and implementation of kinetic effects in MHD models. While with CCMC, she has helped to develop an Open Model Policy, enabling researchers outside the immediate modeling community to have access to modern space science simulations and establishing the CCMC as a leading repository and service center for space weather modeling. Dr. Kuznetsova previously held positions with the Russian Space Research Institute (IKI) and the Raytheon Company, and she currently serves as the chair of the COSPAR Panel on Space Weather and as a liaison to multiple NASA and NSF steering committees. She is a recipient of the NASA Robert H. Goddard Exceptional Achievement Award. Dr. Kuznetsova earned her Ph.D. in theoretical and mathematical physics from the Space Research Institute in Moscow, Russia.

CLIFFORD A. LYNCH is the executive director of the Coalition for Networked Information (CNI). Dr. Lynch is also an adjunct professor at the University of California, Berkeley School of Information. Prior to joining CNI, Dr. Lynch served in the University of California Office of the President, and as director of Library Automation. CNI, jointly sponsored by the Association of Research Libraries and EDUCAUSE, includes about 200 member organizations concerned with the intelligent uses of information technology and networked information to enhance scholarship and intellectual life. CNI's wide-ranging agenda includes work in digital preservation, data-intensive scholarship, teaching, learning and technology, and infrastructure and standards development. Dr. Lynch is both a past president and recipient of the Award of Merit for the American Society for Information Science, and a fellow of the American Association for the Advancement of Science and the National Information Standards Organization. Dr. Lynch earned his Ph.D. in computer science from the University of California, Berkeley. He has served as co-chair of the Board on Research Data and Information, co-chair of the Committee on Planning a Global Library of Mathematical Sciences, and member of the Planning Committee for a Workshop on Overcoming the Technical and Policy Constraints that Limit Large-Scale Data Integration.

MELISSA A. McGRATH is a senior scientist at the SETI Institute. Dr. McGrath's research expertise includes planetary and satellite atmospheres and magnetospheres, particularly imaging and spectroscopic studies of Jupiter's Galilean satellites. She is currently a co-investigator on the Ultraviolet Spectrometer instrument on the European Space Agency (ESA) JUICE mission to Ganymede, as well as a co-investigator on two proposed instruments for NASA's Europa Clipper mission. Previously, Dr. McGrath served as chief scientist at NASA's Marshall Space Flight Center. Dr. McGrath has served as the chair of the American Astronomical Society's Division for Planetary Sciences and as president of the International Astronomical Union's Commission 16 (Physical Studies of Planets and Satellites). She is currently a scientific editor for both the *Astronomical Journal* and the *Astrophysical Journal Letters*. Dr. McGrath has been awarded the NASA Exceptional Service Medal; the NASA Superior Accomplishment Award; and the NASA Ames Honor Award in Lunar Science. Dr. McGrath earned her Ph.D. in astronomy from the University of Virginia.

AARON RIDLEY is a professor at the University of Michigan (UM) in the Department of Climate and Space Science and Engineering. Dr. Ridley previously served as a research scientist at the Southwest Research Institute. His research interests include modeling of the near-Earth space environment, ground-based instrumentation, and small satellites. Dr. Ridley currently has an active program for Fabry-Perot interferometers in North America. He has been PI of three CubeSats, including CADRE and two CubeSats for the European QB50 mission, each of which will measure the state of the upper atmosphere. Dr. Ridley has received the UM's College of Engineering Monroe-Brown Foundation Education Excellence Award, the NASA Group Achievement Award, the UM College of Engineering Outstanding Research Scientist Award, and the Most Cited Paper, *Journal of Atmospheric and Solar-Terrestrial Research*. He earned a B.S. from Eastern Michigan University, and an M.S. and a Ph.D. in atmospheric and space science from the University of Michigan. He has served on the National Academies Committee on Solar and Space Physics and the Committee on Assessment of the National Science Foundation (NSF) 2015 Geospace Portfolio Review.

STAFF

ABIGAIL A. SHEFFER, *Study Director*, is a senior program officer with the Space Studies Board (SSB) of the National Academies. In fall 2009, Dr. Sheffer served as a Christine Mirzayan Science and Technology Policy Graduate Fellow for the National Academies and then joined the SSB. Since joining the National Academies, she has been the staff officer and study director on a variety of activities such as the Committee on Solar and Space Physics, *Assessment of the National Science Foundation's 2015 Geospace Portfolio Review*, *Achieving Science with CubeSats: Thinking Inside the Box*, *Landsat and Beyond—Sustaining and Enhancing the Nation's Land Imaging Program*, among others. Dr. Sheffer has been an assisting staff officer on several other reports, including *Pathways to Exploration—Rationales and Approaches for a U.S. Program of Human Space Exploration* and *Solar and Space Physics: A Science for a Technological Society*. Dr. Sheffer earned her Ph.D. in planetary science from the University of Arizona and A.B. in geosciences from Princeton University.

NATHAN J. BOLL is an associate program officer with the SSB and the Aeronautics and Space Engineering Board (ASEB) of the National Academies. He previously served as a research assistant in civil and commercial space at the Congressional Research Service in the Library of Congress and as a Christine Mirzayan Science and Technology Policy Graduate Fellow at the National Academies. Mr. Boll's background in space policy and science communication includes experience in the Office of International and Interagency Relations at NASA Headquarters, in the Aeronautics and Space Academies at the NASA Glenn Research Center, and as a member of the advisory board of the Montana Space Grant Consortium. Nathan earned his M.S. in space sciences from the University of Michigan, his M.A. in international science and technology policy from George Washington University, and his B.S. in mathematics from the University of Montana Western.

ANESIA WILKS is a senior program assistant. Ms. Wilks began working at the National Academies in the conference management office and later transferred to the Division on Engineering and Physical Sciences (DEPS), where she began working on administrative roles for different projects. She is currently working on the Aeronautics Research and Technology Roundtable and the Space Technology Industry-Government-University Roundtable, among various other projects. Ms. Wilks has a B.A. in psychology (magna cum laude) from Trinity University in Washington, D.C.

CARSON BULLOCK is an undergraduate student in their final year at the College of Wooster. They will receive a B.A. in physics and political science in 2019. Mx. Bullock enjoyed their time as a Lloyd V. Berkner Space Policy Intern during the summer of 2018, a position whose interdisciplinary nature represented a perfect intersection of their interests. Mx. Bollock studies collective action problems and commons management, with an emphasis on the proliferation and mitigation of orbital debris. Outside their major fields of study, Mx. Bullock's broader academic experience includes mathematics, cartography, phonology, and gender.

JONATHAN LUTZ is in his senior year at the University of Colorado Boulder (CU Boulder) in the astrophysics program and worked as a student associate at the Laboratory for Atmospheric and Space Physics. He was the Lloyd V. Berkner Space Policy Intern at the National Academies in autumn 2018. He was a member of a student-led BalloonSat research team that launched a scintillator gamma-ray detector on a small payload to the stratosphere. Previously, he worked as a freelance graphic designer and has a background in data science. He is on the dean's list at CU Boulder.

JACOB ROBERTSON was a Lloyd V. Berkner Space Policy Intern at the SSB during fall 2017. He previously interned with the Education Division at the American Institute of Physics (AIP) and with the Dark Energy Survey at Fermi National Accelerator Laboratory. Mr. Robertson received his B.S. in physics from Austin Peay State University in December 2017. He is currently a program assistant with COMPASS Science Communication, a nonprofit that helps scientists participate in the public dialogue through communication training and by facilitating real-world connections.

E

Acronyms

AAS	American Astronomical Society
ACCESS	Advancing Collaborative Connections for Earth System Science
ACM	Association for Computing Machinery
AGU	American Geophysical Union
AI	artificial intelligence
AIP	American Institute of Physics
AJPS	*American Journal of Political Science*
AMS	American Meteorological Society
AO	Announcement of Opportunity
AON	Arctic Observing Network
API	application programming interface
APL	Applied Physics Laboratory
ARCSS	Arctic System Sciences
ARL	Army Research Laboratory
ASEB	Aeronautics and Space Engineering Board
BIO	Directorate for Biological Sciences
BSD	Berkeley Software Distribution
CAM	Community Atmosphere Model
CC	Creative Commons
CC0	Creative Commons license
CCM	Chemistry Climate Model
CCMC	Community Coordinated Modeling Center
CERN	European Organization for Nuclear Research
CESM	Community Earth System Model
CICE	Community Ice Code
CIO	chief information officer
CISE	Computer and Information Science and Engineering

CLM	Community Land Model
CNI	Coalition for Networked Information
COTS	commercial off-the-shelf
CPU	computer processing unit
DARPA	Defense Advanced Research Projects Agency
DF/BBF	dipolarization fronts and bursty bulk flows
DMP	data management plan
DOD	Department of Defense
DOE	Department of Energy
DOI	digital object identifier
EAR	Export Administration Regulations
EOSDIS	Earth Observing System Data and Information System
EPA	Environmental Protection Agency
ESA	European Space Agency
ESD	Earth Science Division
ESDS	Earth Science Data Systems
FAQ	Frequently Asked Questions
FAR	Federal Acquisition Regulations
FFT	a C subroutine library for Fourier transforms
FIA-NP	Future Internet Architecture-Next Phase
FTE	flux-transfer events
GDP	gross domestic product
GI	guest investigator
GPL	GNU General Public License
GSFC	Goddard Space Flight Center
HAO	High Altitude Observatory
HEASARC	High Energy Astrophysics Science Archive Research Center
ICML	International Conference on Machine Learning
IM	Instructional Memorandum
IP	intellectual property
IPCC	Intergovernmental Panel on Climate Change
ITAR	International Traffic in Arms Regulations
KHI	Kelvin-Helmholtz instability
LAPACK	a linear algebra software library
LGPL	GNU Lesser General Public License
LWS	Living with a Star
MIT	Massachusetts Institute of Technology
ML	machine learning
MPL	Mozilla Public License
NARA	National Archives and Records Administration

NASA	National Aeronautics and Space Administration
NCAR	National Center for Atmospheric Research
NExScI	NASA Exoplanet Science Institute
NFS	NASA FAR Supplement
NOAA	National Oceanic and Atmospheric Administration
NOSA	NASA Open Source Agreement
NPD	NASA Policy Directive
NPG	NASA Procedures and Guidelines
NPR	NASA Procedural Requirements
NSF	National Science Foundation
NumPy	a Python Programming Language Library
OCIO	Office of the Chief Information Officer
OCS	Office of the Chief Scientist
OFAC	Office of Foreign Asset Control
OGC	Office of the General Counsel
OMB	Office of Management and Budget
OS	operating system
OSI	Open Source Initiative
OSS	open source software
OSTP	Office of Science and Technology Policy
PDART	Planetary Data Archiving, Restoration, and Tools
PI	principal investigator
PSD	Planetary Science Division
RoR	runs-on-request
ROSES	Research Opportunities in Space and Earth Sciences
SaaS	software as a service
SaTC	Secure and Trustworthy Cyberspace
scikit-learn	a free python machine learning library
SEADAS	Software design for processing Satellite Data
SMD	Science Mission Directorate
SMEX	Small Explorer
SMP	software management plan
SRA	Software Release Authority
SSB	Space Studies Board
SSC	Scientific Steering Committee
SSE	Software for Science and Engineering
STEM	science, technology, engineering, and mathematics
STMD	Space Technology Mission Directorate
SUA	software user agreement
TESS	Transiting Exoplanet Survey Satellite
THMPROC	THEMIS data processing
TTO	Technology Transfer Office
UCAR	University Corporation for Atmospheric Research
UCLA	University of California, Los Angeles

USGS	U.S. Geological Survey
WACCM	Whole Atmosphere Community Climate Model
WP	white paper